Mathtastic Level 1 Numbers to 10 Teaching Book

Tracy Ashbridge

MEd, Grad Cert, PG Dip, BEd (Hons)

www.Mathtastic.com.au

Mathtastic

Learning Number Step by Step

Tracy Ashbridge

First Edition

2022 Brisbane

© Copyright 2022 Mathtastic: Tracy Ashbridge. All rights reserved

ISBN 978-0-6455822-0-8

The photocopiable pages, number problems booklet and word problems are available at:
http://mathtastic.com.au/mathtastic-resources/
Password: Mathtastic

Contents

How to use Mathtastic 6

Module 1 – adding and subtracting 0,1,2,3 7
    Ideas for teacher lesson (1 hour) 7

Module 2 – add from largest, subtract by counting back 9
    Ideas for teacher lesson (1 hour) 9

Module 3 – rainbow facts 11
    Ideas for teacher lesson (1 hour) 11

Module 4 – add and subtract 10's 13
    Ideas for teacher lesson (1 hour) 13

Module 5 – double and halve 15
    Ideas for teacher lesson (1 hour) 15

Module 6 – near doubles 17
    Ideas for teacher lesson (1 hour) 17

Module 7 – partition by place value 19
    Ideas for teacher lesson (1 hour) 19

Module 8 – add and subtract by compensation 21
    Ideas for teacher lesson (1 hour) 21

Level 1 23

Numbers to 10 Resources 23
    Resources – not included 24
        Lesson Plan 25
        Homework - 15 mins per day 27
        Five Frames – 5 wise and random 29
        Number Track 1-20 31
        Dice +1,2,3 32
        -1,2,3 32
        Blank 100 square 33
        Shut the box strips 34
        Number Grid in 5's 35
        Digit cards and Number Rainbow Cards 36

Ten Frames 5 wise	37
Ten strips	41
Number flips – pairs to 10	42
Subitising cards 1-10	43
Ten frames – pair wise	45
Doubles Bus	49
Doubles and Halves skittles game	50
Blank ten frames	52
Bingo	53
0-3 dice	54
Yahtzee Variation	55
Odd One Out Cards	56
Level 1	**61**
Numbers to 10 Workbook	**61**
Instructions	**63**
Setting out the student book	**64**
Coding the answers	**66**
Module 1	**67**
Draw and solve	67
Draw and Solve	67
Module 2	**68**
Draw and solve	68
Retrieval Practice	68
Module 3	**69**
Draw and solve	69
Retrieval Practice	69
Module 4	**70**
Draw and solve	70
Retrieval Practice	70
Module 5	**71**
Draw and solve	71

    Retrieval Practice 71

Module 6 72

    Draw and solve 72

    Retrieval Practice 72

Module 7 73

    Draw and solve 73

    Retrieval Practice 73

Module 8 – Mixed Review from Module 1 74

    Draw and solve 74

    Draw and solve 74

Level 1 75

Numbers to 10 75

Problem Solving book 75

    Recording Page 77

How to use Mathtastic

Mathtastic can be used to teach 1:1 or small groups. The programs spiral through different levels of numbers, each level addressing the 8 number sense strategies:

1. Add 0,1,2, 3, Subtract 0,1,2, 3
2. Add from largest number by counting on, Subtract by counting back
3. Rainbow facts
4. Adding tens, subtract tens
5. Doubles/ halving
6. Near doubles
7. Partitioning numbers by place value
8. Adding and subtracting by compensating, Bridge 10 with a 9, Bridge 10 with 7 or 8, Round and adjust

There are 8 sections to each lesson.

1. Thinking problems – these are designed to be open ended and challenge the student to think mathematically. This is an opportunity for mathematical discussions and exploration.
2. Subitizing (sub rhymes with cube) – this is the skills of recognising a set of objects without counting and is a key skill which is not always established in students with difficulties in maths.
3. Counting patterns and objects – students need to develop a sense of the number line. This is a skill that students with difficulties are often weak in.
4. Number sense – each session there is a different focus working through the 8 areas of number sense. These are explained and modelled before applying the number sense concept to problems.
5. Game – many students with math difficulties can get anxious about maths and practising the skills through games is a less threatening way to gain the repetition they need. The games have been chosen to specifically practice the skill in focus and also allow for reasoning skills.
6. Word problems – students need to apply their knowledge in problems. These are organised by the 11 different ways of presenting addition and subtraction problems so students don't just learn to solve for the final answer but can be flexible to work around the problem.
7. Number problems – each session there are number problems related to the focus area and spaced retrieval of focus areas.
8. More games

Each module can be used as a lesson or can be split over several lessons depending on the time you have available and the speed the student works through the number sense strategies. The games can be repeated easily, and many have options to extend them.

Module 1 – adding and subtracting 0,1,2,3
Ideas for teacher lesson (1 hour)

Thinking Problems	**Thinking task** Build a picture with Cuisenaire rods – 5 mins
Subitizing	**Subitizing** Five frames - Teach that the strip is 5 – so a whole strip is 5. Show strips, can the child subitize how many they see and also explain how they know? E.g., 4 – I know 5 is a full strip so 4 is one less.
Counting	**Counting – patterns and objects** Count to 20 and back along a number track touch the numbers as you say them Do this a couple of times then start from any number, backwards will need more practice than forwards – look for difficulties 14,13,12,11 Write the numbers from 1-20 on blank 100 square – look at the pattern – colour code the patterns in the tens and the ones. Check for reversal issues – fix as you go and note any which need teaching. 1 2 3 4 5 6 7 8 9 10 11 12 13 14 15 16 17 18 19 20
Number Sense	**Number sense focus area explain, explicit teach and model** **Examples and nonexamples** Use counters to add 1,2,3 and subtract 1,2,3 from a group. Check – Do they know the difference between add and subtract? Teach the symbols + - = Fingers – show me a number, 2 more, 3 less etc – teach language more and less. Write down the number problems you do e.g., 5 fingers, 3 less is 2 so 5-3=2

© Copyright 2022 Mathtastic: Tracy Ashbridge. All rights reserved

Games	**Game/ hands on activity** You need: 1-10 number track and dice labelled +1, +2, +3, -1, -2, -3 Both players start on 5 – roll the dice and follow the dice. Name the number you land on. The winner is the first to "fall off" an end of the number track. Extension – use number track 1-20 and start on 10
Word Problems	**Word Problems** These are all set out as Bett lines – read one line at a time and bet what the question is asking. **Problem Solving Book Level 1** – support with concrete materials – counters and drawing pictures. Track which types of questions the students can and can't do.
Number Problems	**A: Number problems – draw the answer as well as number – pentagon** Model 3+1=4 and 3-1=2, show in the 5 ways • Model - counters • Words – 3 counters plus 1 counter makes 4 counters • Pictures - draw • Equations - 3+1=4 • Contexts – make a number story **B: Retrieval and interleaving practice tasks – Level 1 book**
Games	**Game/ hands on activity** Shut the box game – use a 10-sided dice. Roll the dice – you can shut the box for the number for roll or +/- up to 3. For example: if you roll a 5, you could shut the box on 5 or 6 (+1), 7 (+2) or 8 (+3), 4 (-1), 3 (-2) or 2 (-3). Simplify the game by just adding or subtracting 1 from the number you roll. This develops strategy. Discuss the possible options and which you would use and why. If you don't have the shut the box game, just write out the numbers 1-10 on a strip of paper for each player and cross them out. Game - Roll dice and count the spots – look at the patterns. Using the patterns can you subitize the dots? Which ones can you do? How fast can you see the pattern? Record the numbers you get.

Module 2 – add from largest, subtract by counting back
Ideas for teacher lesson (1 hour)

Thinking Problems	**Thinking task** Odd one out task – 4 pictures, which could be the odd one out – look at different ways of grouping
Subitizing	**Subitizing** Roll dice and count the spots – look at the patterns. Using the patterns can you subitize the dots? Which ones can you do? How fast can you see the pattern?
Counting	**Counting – patterns and objects** Count along a number grid in 5's, each group of 5 on a new row 1 2 3 4 5 6 7 8 9 10
Number Sense	**Number sense focus area explain and model** **Examples and nonexamples** Show 2 numbers (digit cards and objects – object as pile as well as laid out on the table) – which is bigger, smaller, most, least, more, less – check this vocab is understood and related.

	Games	**Game/ hands on activity** Write numbers 1-12 in a line (1 2 3 4 5 6 7 8 9 10 11 12) Roll 3, 6-sided dice. Cross off the numbers in order from 1-12, you can use the numbers on any dice as they are or added together (subtracting 2 dice also allowed).
	Word Problems	**Word Problems** These are all set out as Bett lines – read one line at a time and bet what the question is asking. Problem Solving Book Level 1 – support with concrete materials – counters and drawing pictures. Track which types of questions the students can and can't do.
	Number Problems	**A: Number problems – draw the answer as well as number – pentagon** Model 3+1=4 and 3-1=2, show in the 5 ways - Model - counters - Words – 3 counters plus 1 counter makes 4 counters - Pictures - draw - Equations - 3+1=4 - Contexts – make a number story Teach vertical and horizontal presentation of equations. **B: Retrieval and interleaving practice tasks – Level 1 book**
	Games	**Game/ hands on activity** Shut the box game – use a domines. Put the dominoes in a bag. Pick out a domino - you can shut the box for the number total on the domino or +/- up to 3. For example: if you pick a domino with 5 spots, you could shut the box on 5 or 6 (+1), 7 (+2) or 8 (+3), 4 (-1), 3 (-2) or 2 (-3). This develops strategy. Discuss the possible options and which you would use and why. If you don't have the shut the box game, just write out the numbers 1-12 on a strip of paper for each player and cross them out.

Module 3 – rainbow facts
Ideas for teacher lesson (1 hour)

Thinking Problems	**Thinking task** How many different ways can you show 10 on a reknrek? Record answers as model and equation.
Subitizing	**Subitizing** Introduce a ten frame. Explain it is 2 rows of 5. Show numbers as 5 wise. Introduce 10 strips, the 2 halves are shown differently so you can see where 5 is. Show numbers 5 wise. Show numbers to 10 on fingers five wise – whole hand plus… Can they show in another way?
Counting	**Counting – patterns and objects** Count 1-20 on a bead string, forwards and backwards. Move the beads 1 per number to practice 1-1 correspondence when counting. Watch the pattern from 14-11 when counting backwards.
Number Sense	**Number sense focus area explain and model** **Examples and nonexamples** Show pairs to 10 with numicon and/or Cuisenaire rods. Write out pairs of numbers which equal 10 e.g., 2+8=10 as practice writing equations. Use number flip charts – if there are 10 on the whole picture and I fold it, how many can you see, how many can I see

Games	**Game/ hands on activity** Shut the box 1-10 – use 2 ten-sided dice. Roll the dice, you can choose the pair to 10 from either dice to shut the box. E.g., if you roll an 8 and 5, you could choose to shut the box on 2 (pair to 8) or 5 (pair of 5).	
Word Problems	**Word Problems** These are all set out as Bett lines – read one line at a time and bet what the question is asking. Problem Solving Book Level 1 – support with concrete materials – counters and drawing pictures. Track which types of questions the students can and can't do.	
Number Problems	**A: Number problems – draw the answer as well as number – pentagon** Model 3+1=4 and 3-1=2, show in the 5 ways • Model - counters • Words – 3 counters plus 1 counter makes 4 counters • Pictures - draw • Equations - 3+1=4 • Contexts – make a number story **B: Retrieval and interleaving practice tasks – Level 1 book**	
Games	**Game/ hands on activity** Card game – digits 0-10 or subitizing cards 0-10. Turn over 2 cards, if they make a pair to 10 you can keep them.	

Module 4 – add and subtract 10's
Ideas for teacher lesson (1 hour)

Thinking Problems	**Thinking task** Odd one out task – 4 pictures, which could be the odd one out – look at different ways of grouping
Subitizing	**Subitizing** Introduce Reknrek 1 row = 10, there are 2 groups of 5. Subitize numbers 1-10, 5 wise on 1 row.
Counting	**Counting – patterns and objects** Count 1-20 forwards and backwards using reknrek to practice 1-1 correspondence.
Number Sense	**Number sense focus area explain and model** **Examples and nonexamples** Revisit - Show 2 numbers (digit cards and objects – object as pile as well as laid out on the table) – which is bigger, smaller, most, least, more, less – check this vocab is understood and related.

	Game/ hands on activity
Games	Revisit - Write numbers 1-12 in a line (1 2 3 4 5 6 7 8 9 10 11 12) Roll 3, 6-sided dice. Cross off the numbers in order from 1-12, you can use the numbers on any dice as they are or added together (subtracting 2 dice also allowed).
Word Problems	**Word Problems** These are all set out as Bett lines – read one line at a time and bet what the question is asking. Problem Solving Book Level 1 – support with concrete materials – counters and drawing pictures. Track which types of questions the students can and can't do.
Number Problems	**A: Number problems – draw the answer as well as number – pentagon** Model 3+1=4 and 3-1=2, show in the 5 ways • Model - counters • Words – 3 counters plus 1 counter makes 4 counters • Pictures - draw • Equations - 3+1=4 • Contexts – make a number story Teach vertical and horizontal presentation of equations. **B: Retrieval and interleaving practice tasks – Level 1 book**
Games	**Game/ hands on activity** Revisit - Shut the box game – use dominoes. Put the dominoes in a bag. Pick out a domino - you can shut the box for the number total on the domino or +/- up to 3. For example: if you pick a domino with 5 spots, you could shut the box on 5 or 6 (+1), 7 (+2) or 8 (+3), 4 (-1), 3 (-2) or 2 (-3). This develops strategy. Discuss the possible options and which you would use and why. If you don't have the shut the box game, just write out the numbers 1-12 on a strip of paper for each player and cross them out.

Module 5 – double and halve
Ideas for teacher lesson (1 hour)

Thinking Problems	**Thinking task** Organise sets of objects into pairs – how many pairs can you make?
Subitizing	**Subitizing** Introduce ten frame pairwise (use counters) and reknrek as pair wise, this shows double – student to make doubles on both.
Counting	**Counting – patterns and objects** Count forwards and backwards to 20 in 2's. Use beads in 2's.
Number Sense	**Number sense focus area explain and model** **Examples and nonexamples** Make doubles on the bus, ten frames – which is/ is not a double? Turn over doubles cards 1-5. If 10 on the back what is on the front – do as images and digits. For older students, explain this is the same as 2x table.

Games	**Game/ hands on activity** Doubles or halves skittles game to 10 – roll 6-sided dice, 6 = roll again. Play both doubles and halves versions.
Word Problems	**Word Problems** These are all set out as Bett lines – read one line at a time and bet what the question is asking. Problem Solving Book Level 1 – support with concrete materials – counters and drawing pictures. Track which types of questions the students can and can't do.
Number Problems	**A: Number problems – draw the answer as well as number – pentagon** Model 3+1=4 and 3-1=2, show in the 5 ways • Model - counters • Words – 3 counters plus 1 counter makes 4 counters • Pictures - draw • Equations - 3+1=4 • Contexts – make a number story **B: Retrieval and interleaving practice tasks – Level 1 book**
Games	**Game/ hands on activity** Memory game – use cards 1-5 from a pack of playing cards (20 in total). Shuffle and place the cards face down on the table. Take turns to pick up 3 cards. If you pick up a double, you keep the cards. If not return the cards to the same position on the table.

Module 6 – near doubles
Ideas for teacher lesson (1 hour)

Thinking Problems	**Thinking task** Odd one out task – 4 pictures, which could be the odd one out – look at different ways of grouping
Subitizing	**Subitizing** Mix of ten frames pairwise, reknrek pairwise, dominoes pairwise, fingers pairwise, numicon, Cuisenaire (doubles facts)
Counting	**Counting – patterns and objects** Count forwards and backwards to 20 in 2's along reknrek.
Number Sense	**Number sense focus area explain and model** **Examples and nonexamples** Explain near double, make near doubles on ten frames and reknrek. Number flips if you know double 5, what are the near doubles.

Games	**Game/ hands on activity** Card war – pick 2 cards - you may keep if it is a double or near double.
Word Problems	**Word Problems** These are all set out as Bett lines – read one line at a time and bet what the question is asking. Problem Solving Book Level 1 – support with concrete materials – counters and drawing pictures. Track which types of questions the students can and can't do.
Number Problems	**A: Number problems – draw the answer as well as number – pentagon** Model 3+1=4 and 3-1=2, show in the 5 ways • Model - counters • Words – 3 counters plus 1 counter makes 4 counters • Pictures - draw • Equations - 3+1=4 • Contexts – make a number story **B: Retrieval and interleaving practice tasks – Level 1 book**
Games	**Game/ hands on activity** Near doubles skittles game – use numbers 1,3,5,7,9,11. Roll the dice (6= roll again). Roll the dice and cover the near double.

Module 7 – partition by place value
Ideas for teacher lesson (1 hour)

Thinking Problems	**Thinking task** Odd one out task – 4 pictures, which could be the odd one out – look at different ways of grouping
Subitizing	**Subitizing** Dominoes – look at the dominoes – what numbers can you see on each side (don't look at the total of the whole domino yet). Reteach patterns to 6 from dice activity last week. Introduce tally marks – review 5 and some more. Subitize to 10.
Counting	**Counting – patterns and objects** Count to 11-20 using reknrek. 11= 1 ten and 1 one, 12= 1 ten and 2 ones etc Write out the pattern 11-20 forwards and backwards – what do they notice?
Number Sense	**Number sense focus area explain and model** **Examples and nonexamples** Discuss how numbers are made of 5 plus ? Up to 10. Write and draw equations for these.

Games	**Game/ hands on activity** Bingo – 0-3 dice. Write numbers from 5-8 into the squares. Roll the dice and cover the number which is 5 plus the number.
Word Problems	**Word Problems** These are all set out as Bett lines – read one line at a time and bet what the question is asking. Problem Solving Book Level 1 – support with concrete materials – counters and drawing pictures. For this module finish any outstanding problems. Track which types of questions the students can and can't do.
Number Problems	**A: Number problems – draw the answer as well as number – pentagon** Model 3+1=4 and 3-1=2, show in the 5 ways • Model - counters • Words – 3 counters plus 1 counter makes 4 counters • Pictures - draw • Equations - 3+1=4 • Contexts – make a number story **B: Retrieval and interleaving practice tasks – Level 1 book**
Games	**Game/ hands on activity** Use playing cards 1-5. Take turns to turn over 2 cards and add them together. Can you reorganise them to be 5+? For example: if you turn over 2 and 4 that makes 6 which is the same as 1+5

Module 8 – add and subtract by compensation
Ideas for teacher lesson (1 hour)

Thinking Problems	**Thinking task** Odd one out task – 4 pictures, which could be the odd one out – look at different ways of grouping
Subitizing	**Subitizing** Review ten frames 5 wise and pair wise. Subitize and 5 and ? Ten strips. Subitize and 5 and ? Reknrek. Subitize and 5 and ? Fingers Subitize and 5 and ? Tally marks – subitize 5 and ?
Counting	**Counting – patterns and objects** Count to 20 using number rainbow. Put out rainbow in order and the point and count forward and backwards to 20.
Number Sense	**Number sense focus area explain and model** **Examples and nonexamples** Teach that 4+2 could be rearranged as 5+1. Find all the different ways this could be organised. Use Cuisenaire rods to show visually.

Games	**Game/ hands on activity** Yahtzee variation - with 3 dice Odd/even Near doubles to 5 Doubles +/- 1,2,3
Word Problems	**Word Problems** These are all set out as Bett lines – read one line at a time and bet what the question is asking. Problem Solving Book Level 1 – support with concrete materials – counters and drawing pictures. For this module finish any outstanding problems. Track which types of questions the students can and can't do.
Number Problems	**A: Number problems – draw the answer as well as number – pentagon** Model 3+1=4 and 3-1=2, show in the 5 ways - Model - counters - Words – 3 counters plus 1 counter makes 4 counters - Pictures - draw - Equations - 3+1=4 - Contexts – make a number story **B: Retrieval and interleaving practice tasks – Level 1 book**
Games	**Game/ hands on activity** Domino snap to 10 – you can snap the 2 dominoes if they add to 10 Card snap – as above – you can snap the cards if they equal 10 (remove 10 and picture cards, explain that ace =1)

Level 1
Numbers to 10
Resources

Resources – not included

Reknrek 1-20

Dice 1-6

Dice 1-10

Blank 6-sided dice

Bead string 1-20 (in 2's and 5's)

Cuisenaire rods

Numicon – optional

Counters

Shut the box game – optional – paper version as alternative

Dominoes

Playing cards

Lesson Plan

	Thinking task
Thinking Problems	
Subitizing	**Subitizing**
Counting	**Counting – patterns and objects**
Number Sense	**Number sense focus area explain and model** **Examples and nonexamples**
Games	**Game/ hands on activity**

Word Problems	**Word Problems** These are all set out as Bett lines – read one line at a time and bet what the question is asking. Problem Solving Book Level 1 – support with concrete materials – counters and drawing pictures. Track which types of questions the students can and can't do.
Number Problems	**A: Number problems – draw the answer as well as number – pentagon** Model 3+1=4 and 3-1=2, show in the 5 ways • Model - counters • Words – 3 counters plus 1 counter makes 4 counters • Pictures - draw • Equations - 3+1=4 • Contexts – make a number story **B: Retrieval and interleaving practice tasks – Level 1 book**
Games	**Game/ hands on activity**

Homework - 15 mins per day

Day 1	
Counting	
Number Problems	
Games	
Day 2	
Counting	
Problem Solving	Word Problem solving booklet - Level 1
Games	

Day 3	
Counting	
Number Problems	Number problems booklet Level 1– draw the answer as well as the calculation
Games	
Day 4	
Counting	
Problem Solving	Word Problem solving booklet - Level 1
Games	

Five Frames – 5 wise and random

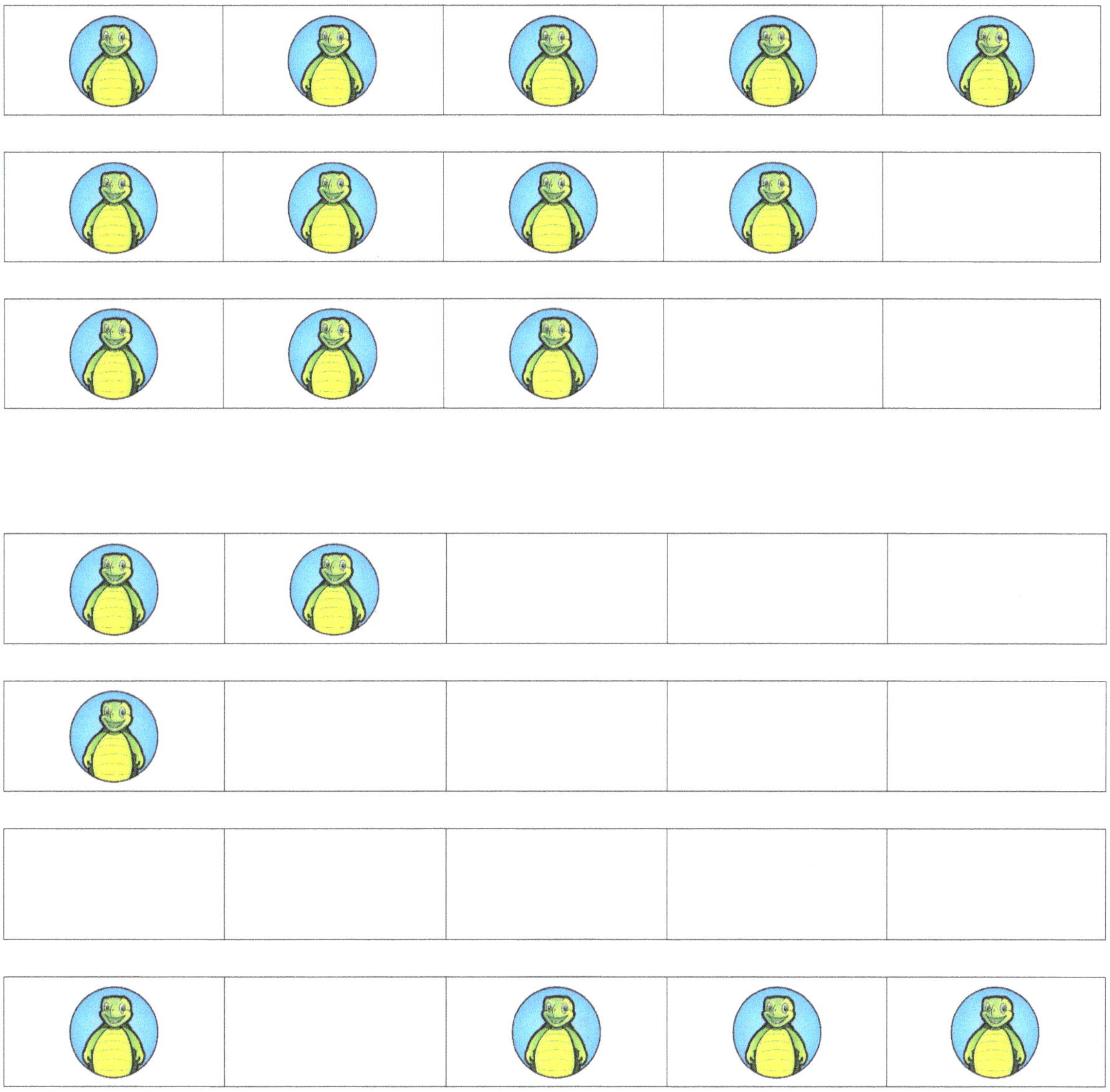

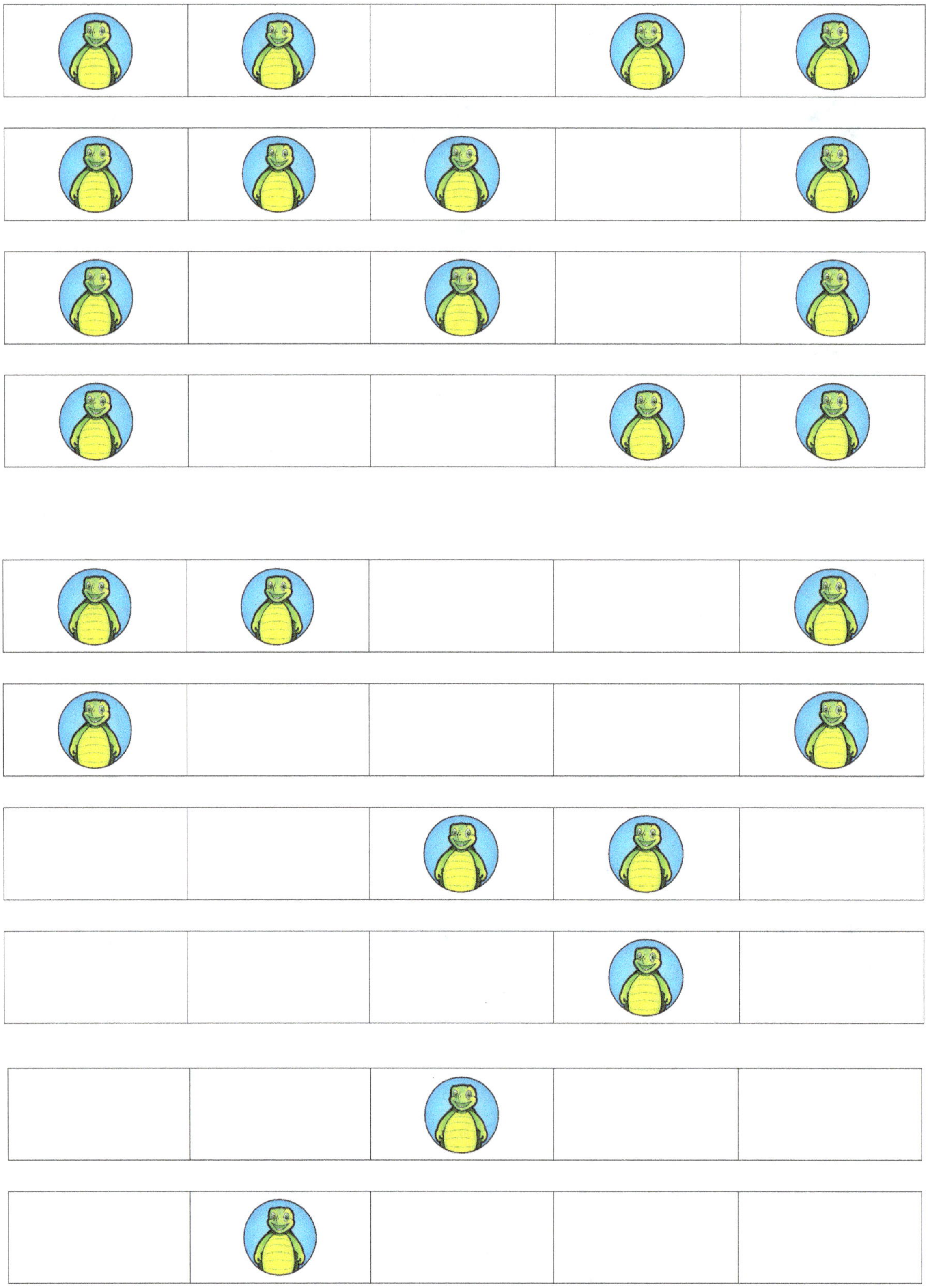

🐢	🐢		🐢	🐢

🐢	🐢	🐢		🐢

🐢		🐢		🐢

🐢			🐢	🐢

🐢	🐢			🐢

🐢				🐢

		🐢	🐢	

			🐢	

		🐢		

	🐢			

Number Track 1-20

1	2	3	4	5	glue
6	7	8	9	10	glue
11	12	13	14	15	glue
16	17	18	19	20	Cut off

Dice
+1,2,3
-1,2,3

	+1		
-2	-1	+2	+3
	-3		

Blank 100 square

Shut the box strips

1	2	3	4	5	6	7	8	9	10
1	2	3	4	5	6	7	8	9	10
1	2	3	4	5	6	7	8	9	10
1	2	3	4	5	6	7	8	9	10
1	2	3	4	5	6	7	8	9	10
1	2	3	4	5	6	7	8	9	10
1	2	3	4	5	6	7	8	9	10

Number Grid in 5's

1	2	3	4	5
6	7	8	9	10
11	12	13	14	15
16	17	18	19	20

Digit cards and Number Rainbow Cards

1	2	3	4	5
6	7	8	9	10
11	12	13	14	15
16	17	18	19	20
0				

Ten Frames 5 wise

Ten strips

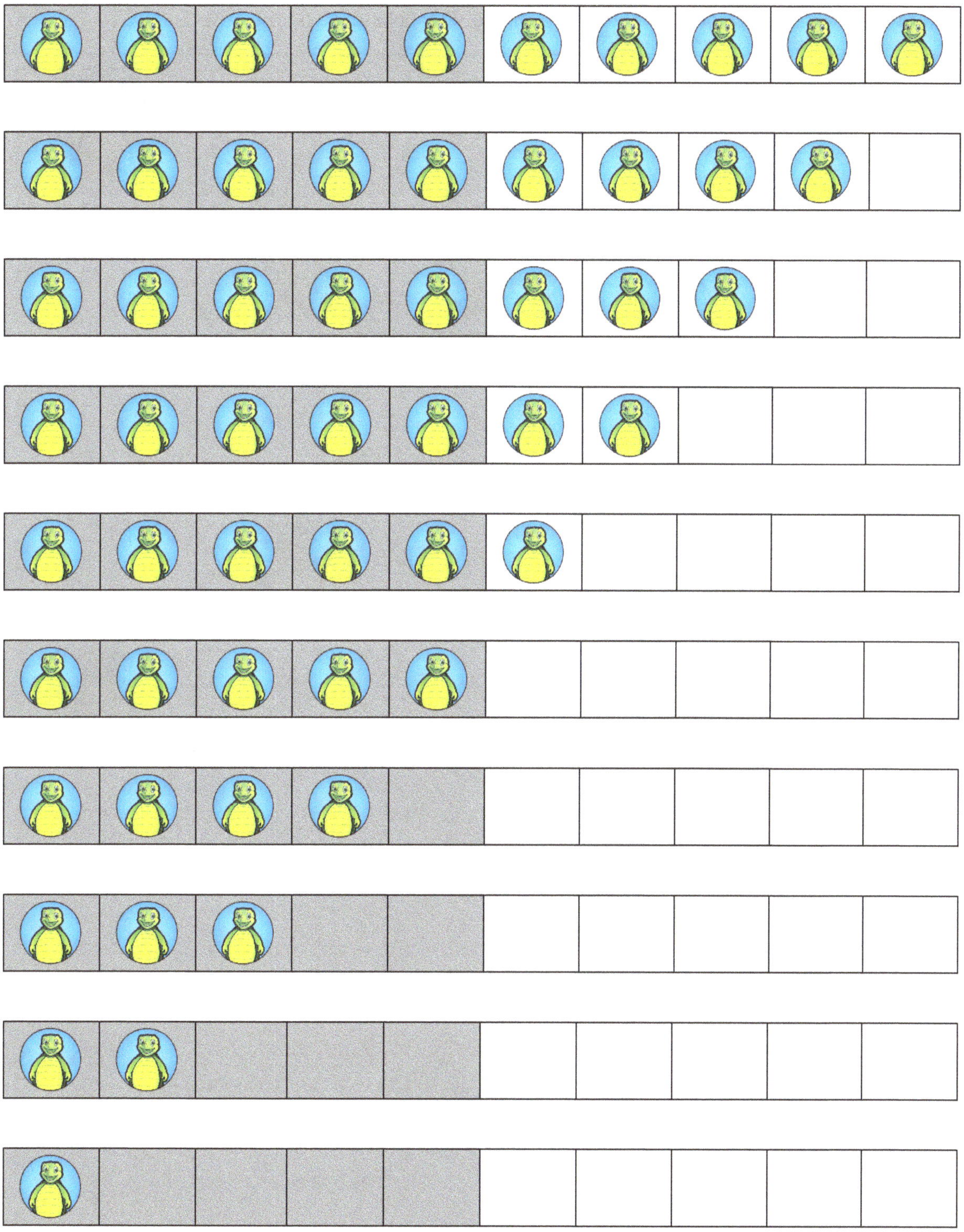

Number flips – pairs to 10

Fold to show different amounts – how many on one side, how many on the other?

🐢				🐢
	🐢	🐢	🐢	
	🐢			
🐢			🐢	
	🐢			🐢

Subitising cards 1-10

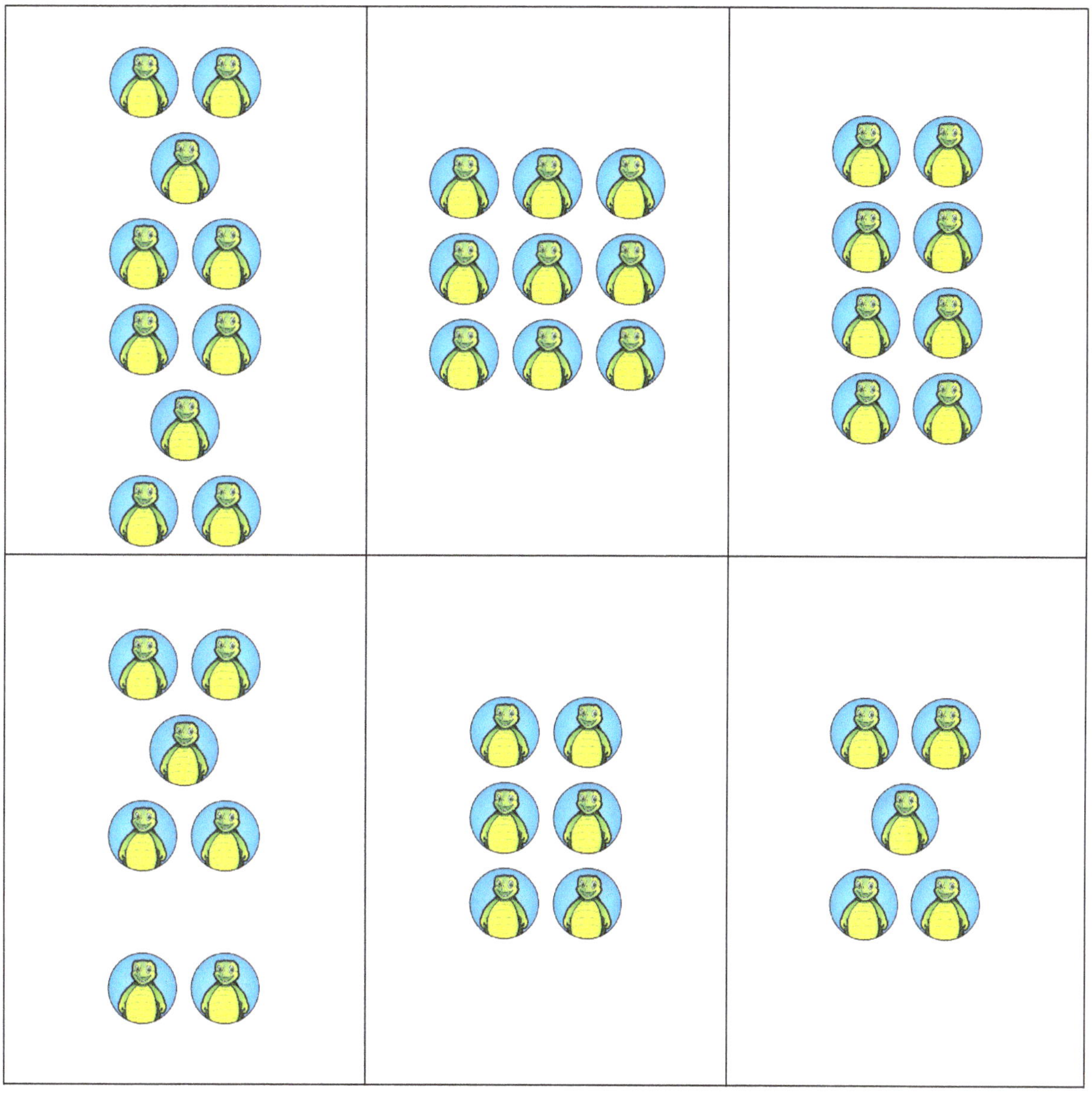

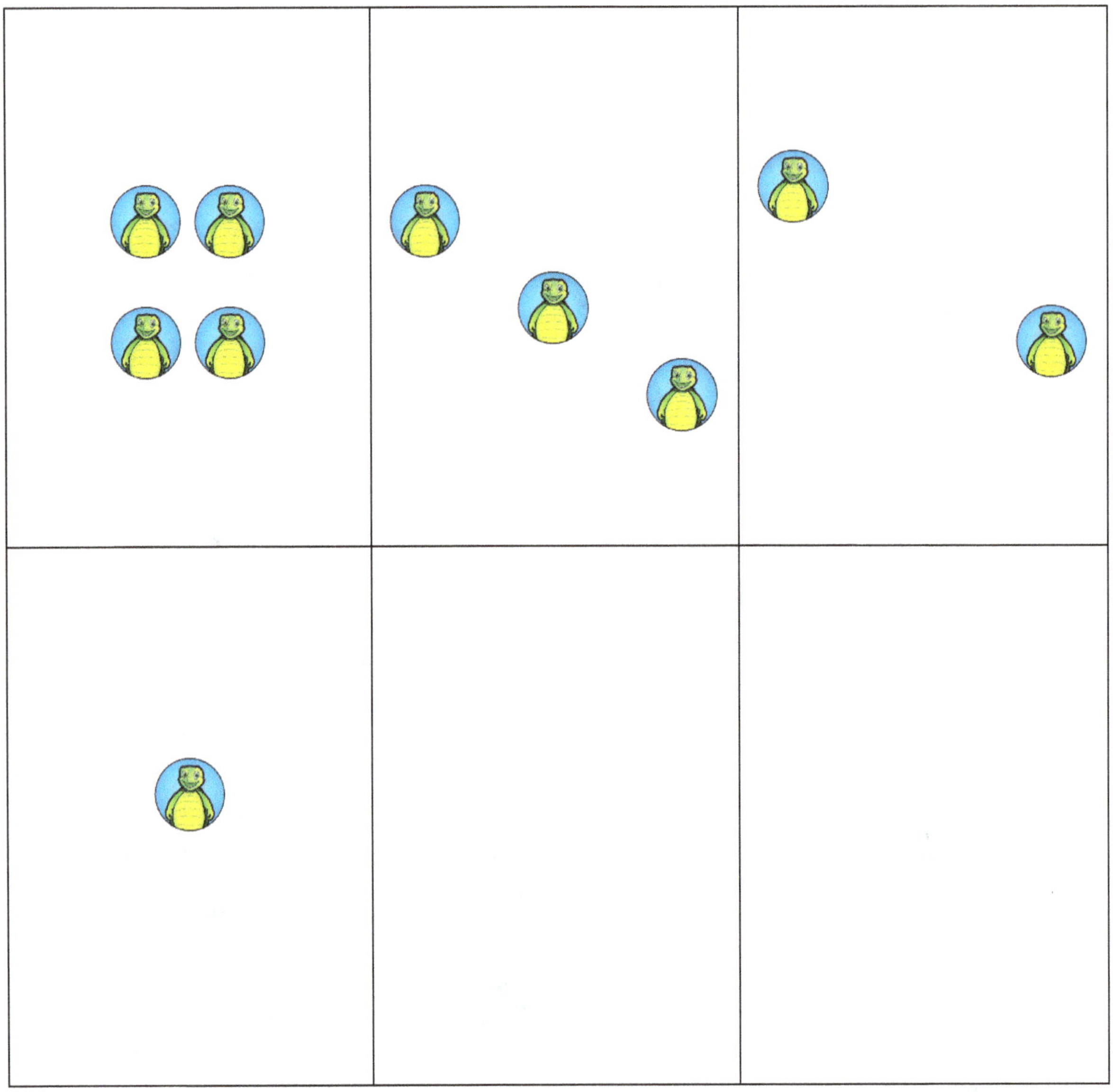

Ten frames – pair wise

Doubles Bus

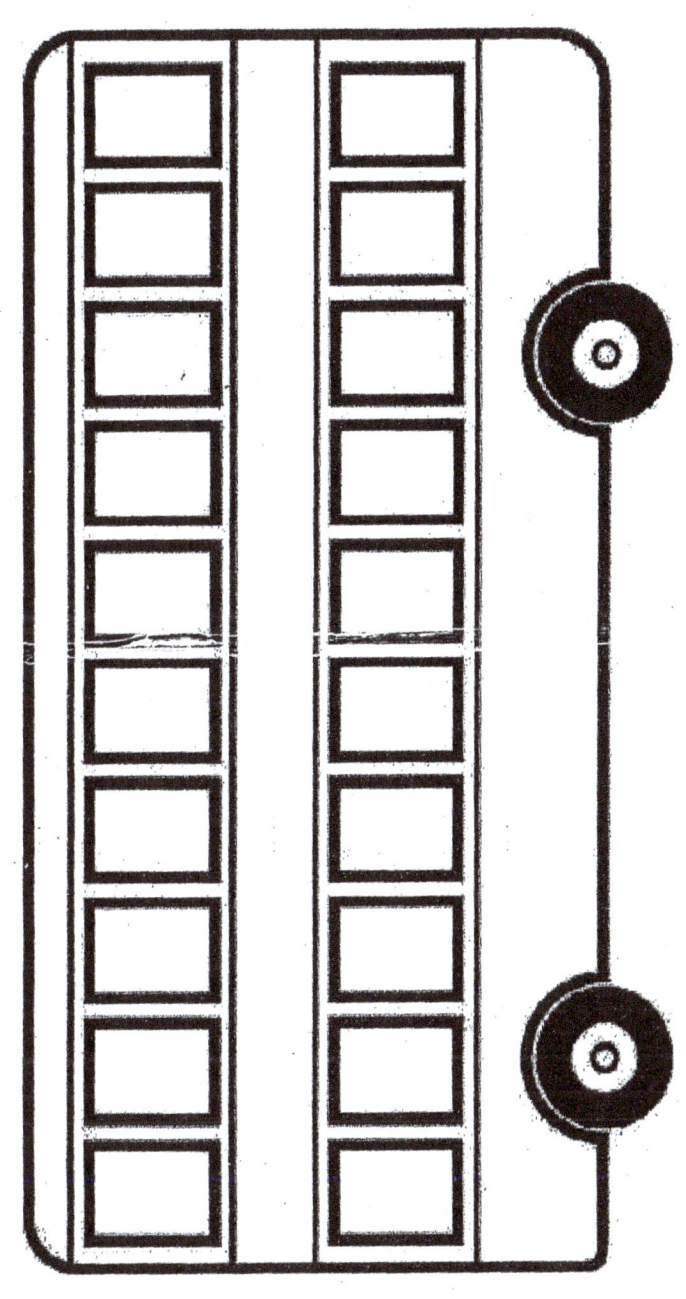

Doubles and Halves skittles game

How to play:

Doubles: write 9 even numbers from 2-10 on the skittles for the doubles game. Roll a 6-sided dice and cross off the double.

Halves: Write 9 numbers from 1-5 on the skittles for halves games. Roll 6-sided dice labelled even numbers 2-10 and *, roll the dice and cross off half.

Near Doubles: write 9 numbers from 1,3,5,7,9,11. Roll a 6-sided dice and cross of a near double e.g., roll 3 can cross off 5 or 7 (double 3 is 6 so near double is 5 or 7)

Winner is the first to cross off all skittles.

If you roll a 6 or *, roll again.

Miss your go if you cannot cross off a skittle.

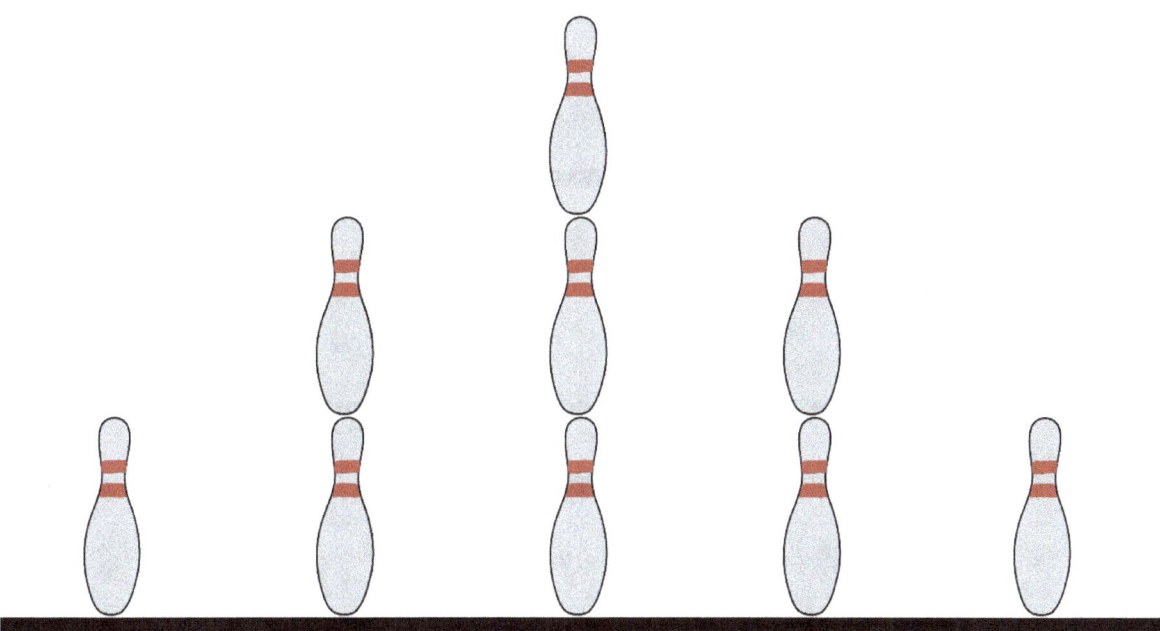
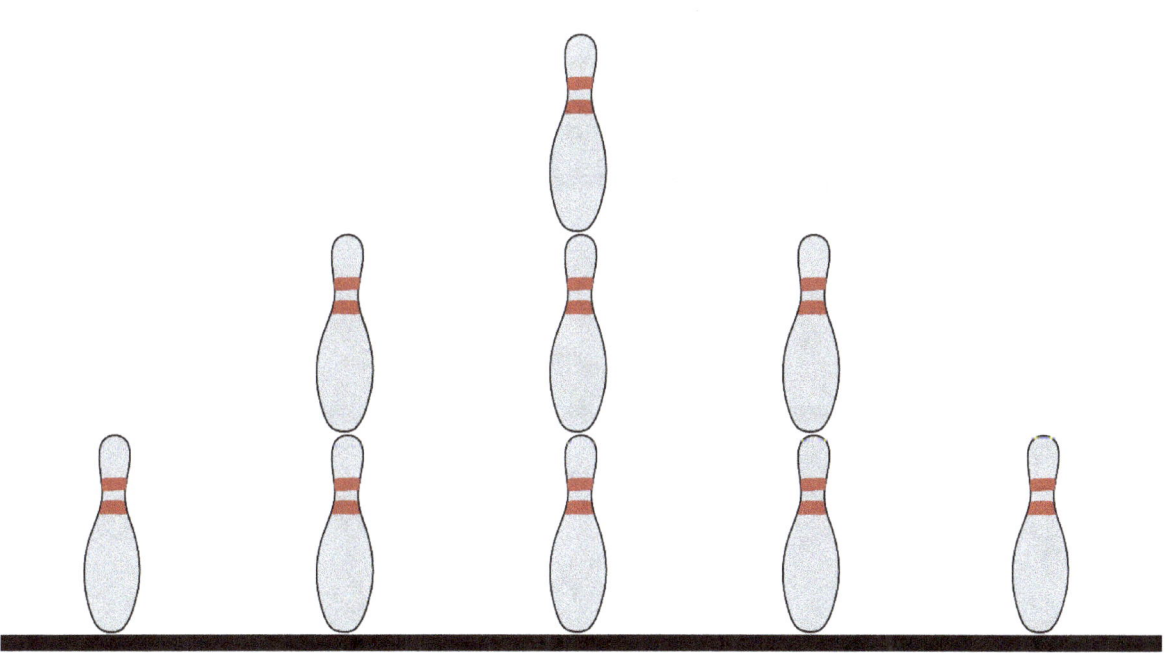

Blank ten frames

Bingo

You will need a 0-3 dice. Write numbers from 5-8 into the squares. Roll the dice and cover the number which is 5 plus the number.

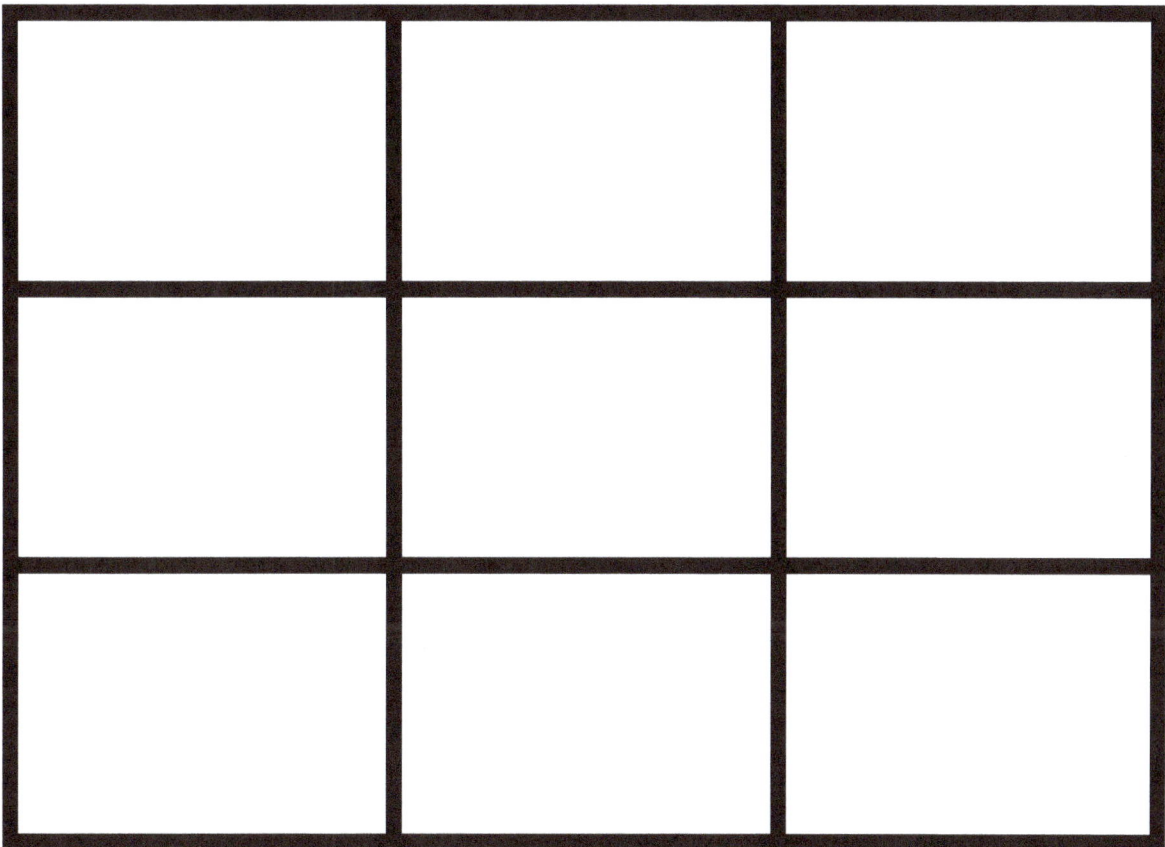

0-3 dice

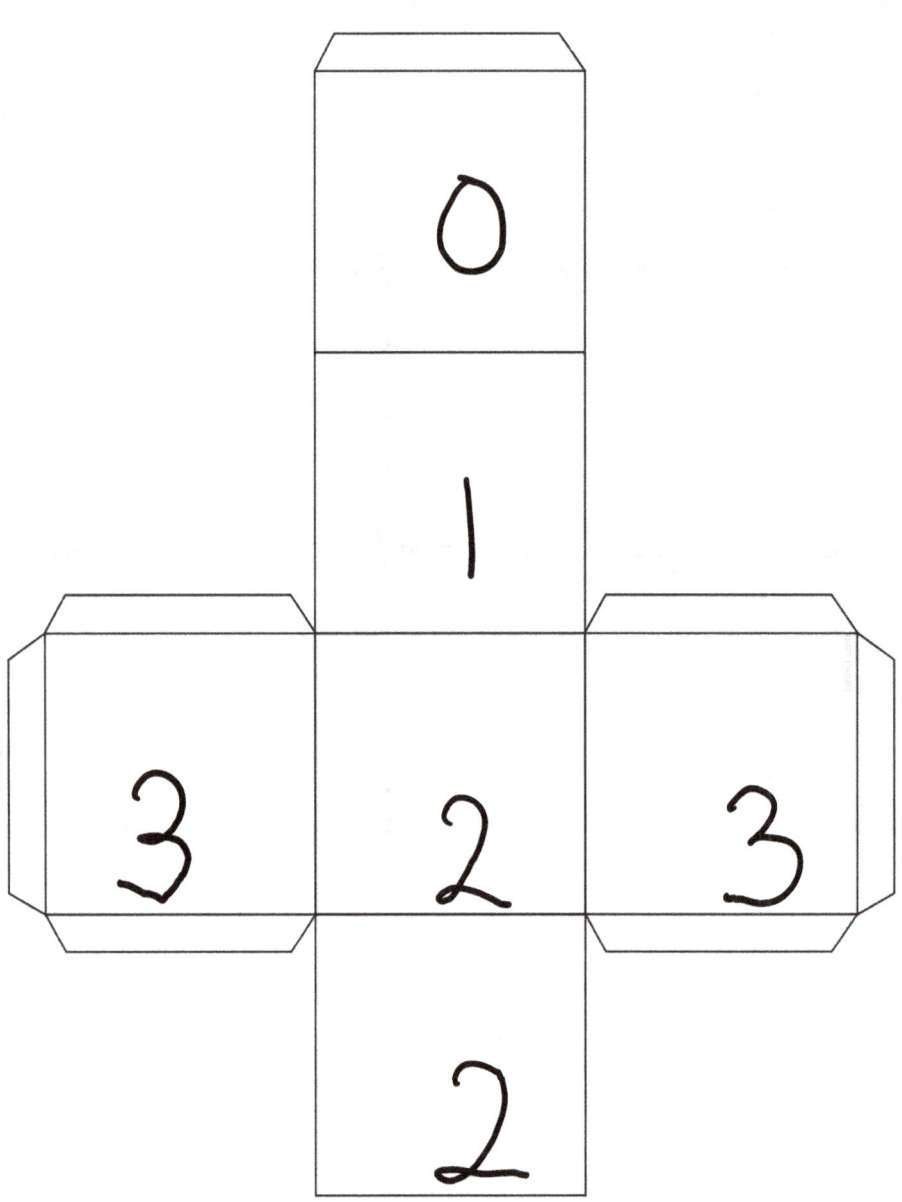

Yahtzee Variation

Use 3 dice, you may roll three times.

	Points	Game 1	Game 2	Game 3
3 odd numbers	10			
3 even numbers	10			
Triple	30			
Double – any 2 of the dice	20			
Near double e.g., 3 and 4	20			
+2 – one dice plus 2 = another dice	10			
+3 - one dice plus 3 = another dice	10			

Odd One Out Cards

Which is the odd one out and why? There is more than 1 answer and reason.

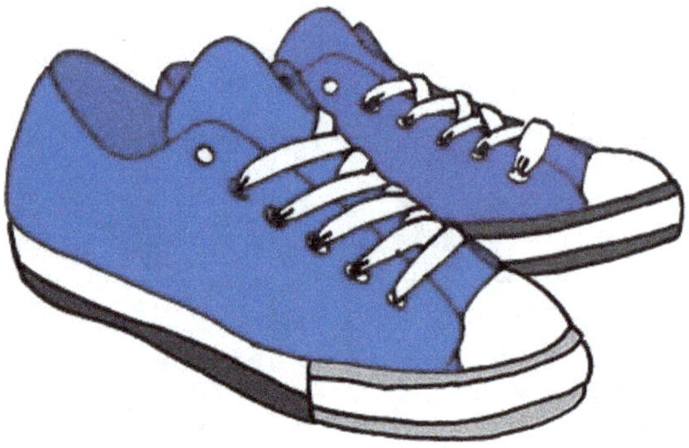

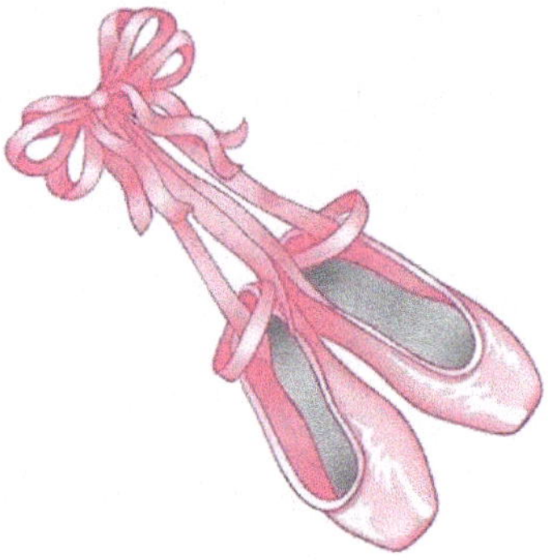

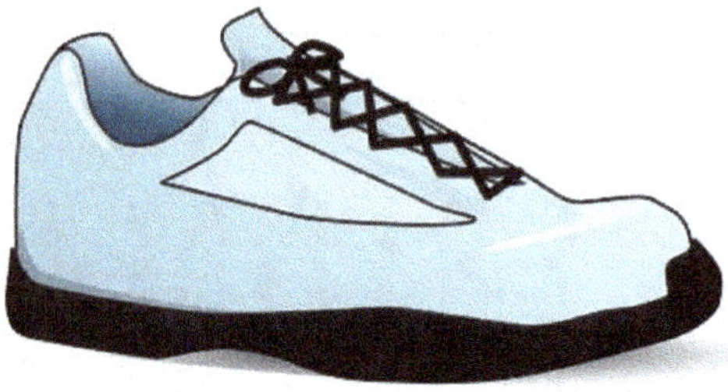

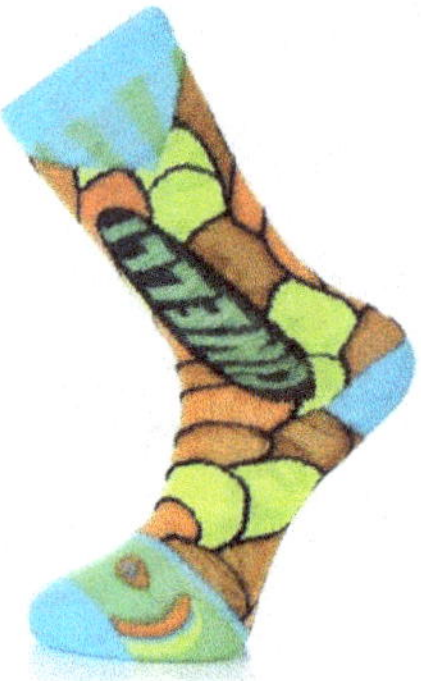

Which is the odd one out and why? There is more than 1 answer and reason.

Which is the odd one out and why? There is more than 1 answer and reason.

Which is the odd one out and why? There is more than 1 answer and reason.

Which is the odd one out and why? There is more than 1 answer and reason.

Level 1

Numbers to 10 Workbook

© Copyright 2022 Mathtastic: Tracy Ashbridge. All rights reserved

Contents

- Instructions 63
- Setting out the student book 64
- Coding the answers 66
- Module 1 67
  - Draw and solve 67
  - Draw and Solve 67
- Module 2 68
  - Draw and solve 68
  - Retrieval Practice 68
- Module 3 69
  - Draw and solve 69
  - Retrieval Practice 69
- Module 4 70
  - Draw and solve 70
  - Retrieval Practice 70
- Module 5 71
  - Draw and solve 71
  - Retrieval Practice 71
- Module 6 72
  - Draw and solve 72
  - Retrieval Practice 72
- Module 7 73
  - Draw and solve 73
  - Retrieval Practice 73
- Module 8 – Mixed Review from Module 1 74
  - Draw and solve 74
  - Draw and solve 74

© Copyright 2022 Mathtastic: Tracy Ashbridge. All rights reserved

Instructions

Answer the questions on each page. Each row is for a different day. The first column contains practice related to the work in the module and the second column is for practice from previous modules. Problems should be presented alternatively horizontally and vertically from Module 2 onwards.

Students (or adults) should copy the problem into a maths exercise book – this can be plain, lined or squared. In the early stages a plain book may be easier.

Divide the page into 2 columns – the first for writing the problem and the second to represent in the 5 different ways. See below for an example.

For each of the questions in the first column solve the problem and note how it was solved in the third column – see below. Then show understanding using one or more of the ways shown on the conceptual understanding pentagon below.

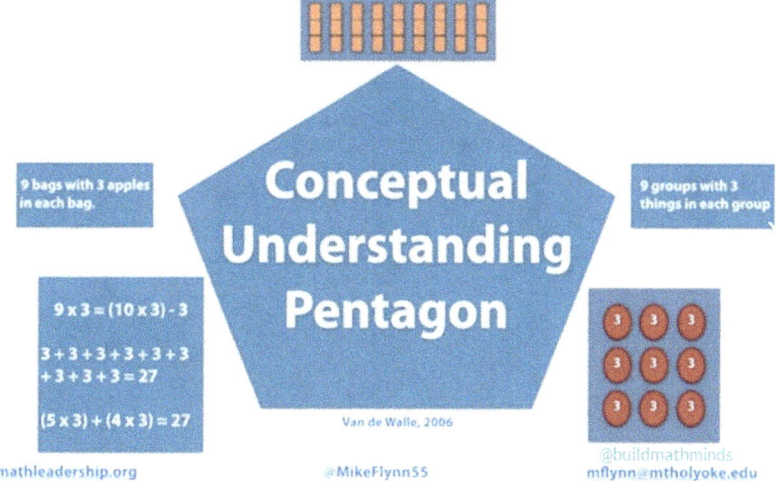

This way students can show their conceptual understanding or misunderstanding:

- Model – show on a model e.g., ten frame, reknrek, numicon
- Words – write out in works – six plus three equals nine.
- Pictures – draw out the problem as a picture
- Equations – write the equation (already done for you but you could rewrite vertically or horizontally)
- Contexts – write a number story – e.g. I had 3 apples, and I ate 1. Now there are only 2 apples.

© Copyright 2022 Mathtastic: Tracy Ashbridge. All rights reserved

Setting out the student book

$6 + 2 = 8$

Six plus two equals eight.

Jane had 6 puppies and her Poppy brought her 2 more. Now she has 8 puppies!

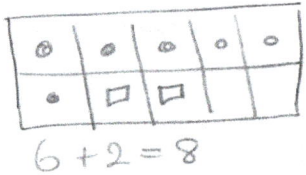

$6 + 2 = 8$

$4 - 1 = 3$

Four subtract one equals three.

Four birds were in the garden. One flew away. Now there are three birds

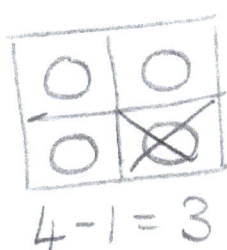

$4 - 1 = 3$

Coding the answers

Coding the answers will help to diagnose which strategies the students are using and which they are not. We really need students to move beyond counting in ones and seeing numbers in bigger groups as well as learning some of these facts so they can recall them automatically.

Coding the student answer strategy: ask the student how they worked out the answer – they may use more than one way	
Automatic – just knew the answer, immediate recall	A
Counted on or back a small number (+/- 0,1,2,3)	S
Rainbow facts – knew the pair made 10	RF
Counted on from largest number	CO
Counted back for subtraction	CB
Doubled or halved	D or H
Near double	ND
Place value – e.g., 10+4= 14	PV
Compensated – made an adjustment to the number before calculating e.g., 9+6 – rearranged to 10+5 as easier	C
Other – you may wish to note this down	O

Module 1

Draw and solve	Code	Draw and Solve	Code
9+0= 4-1 = 6+2 = 7-2= 5+3=		No retrieval this module – more practice questions instead.	
8-1= 4+2= 7-0= 10-2= 6+3=		1+3= 2+3= 9-2= 8-0= 7-3=	
5-0= 4+3= 5+2= 6-2= 7-1=		3-2= 7+3= 8+0= 10-3= 9+1=	

Module 2

Draw and solve	Code	Retrieval Practice	Code
9+1= 2+8= 9-4= 7-3=		7-0= 8+1= 5-3= 7+3= 8-1=	
2+7= 5+1= 8-5= 6-4=		6-1= 7+2= 7-1= 5+2=	
6+3= 1+8= 10-7= 5-3=		5-2= 5+3= 6+3= 8-2= 7-2=	
1+6= 2+4= 8-4= 10-4=		4-3= 5+0= 9-3= 8-3= 8+2=	

Module 3

Draw and solve	Code	Retrieval Practice	Code
Solve – is it a rainbow fact? Find the odd one out. 10+0= 5+5= 8+2= 7+1=		4+1= 6-2= 8-5= 2+5=	
7+3= 6+2= 9+1= 5+5=		8-3= 5+3= 7-5= 5+4=	
2+8= 3+7= 4+6= 9+2=		4+2= 7-0= 9-6= 1+8=	
1+6= 5+5= 1+9= 6+4=		8-1= 9+0= 8-6= 2+7=	

Module 4

Draw and solve	Code	Retrieval Practice	Code
9-5= 2+5= 10-5= 5+4=		6+3= 7+3= 10-4= 2+6= 6-5=	
8-5= 3+5= 10-6= 6-5=		9-3= 8+2= 10-5= 4+5= 10-7=	
7-5= 4+5= 9-5= 5+3=		7+2= 6+4= 10-6= 1+8= 8-7=	
6-5= 1+5= 10-4= 5+5=		10-2= 9+1= 10-8= 0+10= 7-4=	

Module 5

Draw and solve	Code	Retrieval Practice	code
5+5= 4+4= 3+3= 1+1=		7+1= 7-6= 8+2= 7-5=	
0+0= 4-2= 10-5= 6-3=		8-2= 2+6= 7+3= 5+3=	
8-4= 2+2= 2-1= 4-2=		9+0= 1+7= 6+4= 9-5=	
3+3= 0+0= 10-5= 4-2=		6+3= 8-5= 5+5= 10-5=	

Module 6

Draw and solve	Code	Retrieval Practice	code
5+4= 3+2= 1+2= 4+3=		6+2= 7+3= 8-5= 2+2= 10-5=	
2+3= 4+5= 2+1= 1+0=		9-7= 4+6= 9-6= 8-4= 3+3=	
4+3= 3+2= 1+2= 5+4=		7-3= 5+5= 5-5= 6-3= 4+4=	
2+3= 4+5= 2+1= 1+2=		3+6= 1+9= 7-5= 4-2= 5+5=	

Module 7

Draw and solve	Code	Retrieval Practice	Code
The focus of this module is looking at groups of 5 and ? 5+1= 9-5=		6+0= 10-7= 4+4= 4+5= 2+2= 2+3=	
5+2= 8-5=		1+6= 5+5= 5+4= 8-5= 6-3= 4+3=	
5+3= 7-5=		3+3= 3+4= 7-2= 10-4= 8-4= 3+2=	
5+4= 6-5=		4+4= 3+4= 9-7= 5+2= 5+5= 8-4=	

© Copyright 2022 Mathtastic: Tracy Ashbridge. All rights reserved

Module 8 – Mixed Review from Module 1

Draw and solve	Code	Draw and solve	Code
6+2= 2-1= 4-4= 3+6= 5+5= 5-2= 7-6= 10-8=		2+4= 1+3= 7+3= 9-7= 3-1= 9-8= 8-2= 5+3=	
4+6= 4-2= 1-1= 8-7= 8+2= 2+5= 4+4= 4+5=		3+3= 3+4= 4-3= 8-6= 2-2= 8-1= 6+1= 9-2=	

Level 1
Numbers to 10
Problem Solving book

How to use these cards

How to use these problem cards

Each card is written to meet each of the different ways of presenting a problem. These are written at the bottom of each page for teacher reference.

The problems are deliberately written with each piece of the problem written on a separate line. This way you can cover up the rest of the problem and reveal one line at a time.

Bet lines- this is a technique to help students to think about the problem. You show one line at a time and as students to "bet" what the problem is going to ask them to do. They review their ideas as more information is revealed.

For each problem there are 2 extra sets of numbers which can be used instead for extra practice of that problem type -at the bottom of the page in brackets next to the problem type)

Recording Page

Use this page to record how the student managed with each different problem type. Which can they do easily, and which need more practice?

Join – result unknown 6+2=?	Join – change unknown 6+?=8	Join – start unknown ?+2=8
Separate – result unknown 9-5=?	Separate – change unknown 9-?=4	Separate – start unknown ?-5=4
Part-part-whole – whole unknown 5+4=?	Part- part-whole – part unknown ?+4=9	
Compare – difference unknown 7-2=?	Compare – compared set unknown 7-?=5	Compare – referent unknown ?-5=2

© Copyright 2022 Mathtastic: Tracy Ashbridge. All rights reserved

Mathstastic
CGI Math Problems

Mathtastic Level 1 - Numbers to 10

Add or subtract a small number

tracyashbridge.com

Georgie had 5 crayons.

Sam gave her 2 more for her birthday.

How many crayons does she have now?

Join - result unknown (7/1, 8/3)
Mathtastic Level 1 - Numbers to 10
Add or subtract a small number
tracyashbridge.com

Georgie had 6 crayons.

She got some more for her birthday.

Now she has 8 crayons.

How many crayons did she get for her birthday?

Join - change unknown (5/7, 6/9)
Mathtastic Level 1 - Numbers to 10
Add or subtract a small number
tracyashbridge.com

© Copyright 2022 Mathtastic: Tracy Ashbridge. All rights reserved

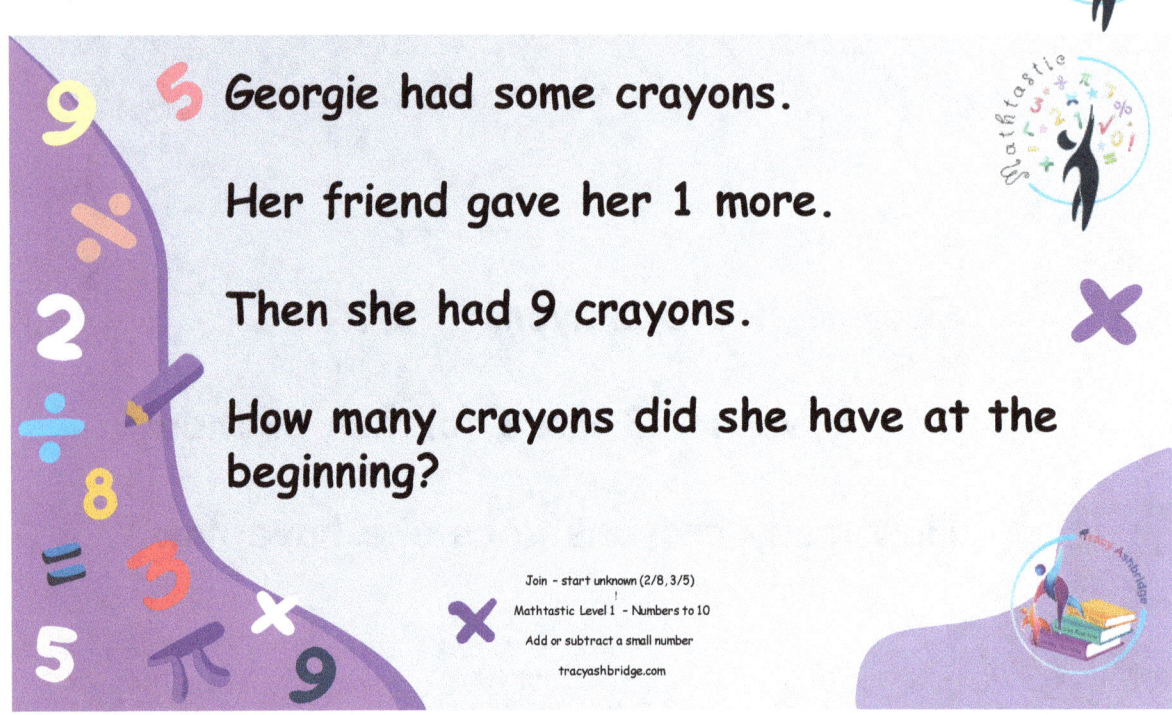

Georgie had some crayons.

Her friend gave her 1 more.

Then she had 9 crayons.

How many crayons did she have at the beginning?

Join – start unknown (2/8, 3/5)
Mathtastic Level 1 – Numbers to 10
Add or subtract a small number
tracyashbridge.com

Sam had 5 pencils.

She gave 1 to Sarah.

How many pencils does Sam have left?

Separate – result unknown (7/2, 8/3)
Mathtastic Level 1 – Numbers to 10
Add or subtract a small number
tracyashbridge.com

Georgie had 7 crayons.

She gave some to Tracy.

Georgie has 6 crayons left?

How many did she give to Tracy?

Separate – change unknown (8/2, 5/4)
Mathtastic Level 1 – Numbers to 10
Add or subtract a small number
tracyashbridge.com

Georgie had some crayons.

She gave 2 to Sarah.

Then she had 7 left.

How many crayons did Georgie have to start with?

Separate – start unknown (3/5, 2/8)
Mathtastic Level 1 – Numbers to 10
Add or subtract a small number
tracyashbridge.com

© Copyright 2022 Mathtastic: Tracy Ashbridge. All rights reserved

4 green crayons and 2 blue crayons were in the pot.

How many were in the pot altogether?

Part – Part– Whole – whole unknown (6/3, 8/1)
Mathtastic Level 1 – Numbers to 10
Add or subtract a small number
tracyashbridge.com

There were 6 crayons. Some were red and some were yellow.

4 were red.

How many were yellow?

Part – Part– Whole – part unknown (8/2, 7/5)
Mathtastic Level 1 – Numbers to 10
Add or subtract a small number
tracyashbridge.com

Georgie had 6 crayons.

Jane has 4 crayons.

Georgie has how many more crayons than Jane?

Compare – difference unknown (7/4, 5/3)
Mathtastic Level 1 – Numbers to 10
Add or subtract a small number
tracyashbridge.com

Georgie has 3 crayons.

Sam has 2 more crayons than Georgie.

How many crayons does Sam have?

Compare – compared set unknown
Mathtastic Level 1 – Numbers to 10
Add or subtract a small number
tracyashbridge.com

Mathstastic
CGI Math Problems

Mathstastic Level 1 – Numbers to 10

Add from largest, subtract by counting back to 10

tracyashbridge.com

Gemma had 2 biscuits.
Then she got 7 more.
How many biscuits does she have now?

Join - result unknown (3/5, 6/2)
Mathtastic Level 1 - Numbers to 10
Add from largest, subtract by counting back
tracyashbridge.com

Jim had 6 lollies.
He bought some more.
Now he has 9 lollies.
How many lollies did he buy?

Join - change unknown (5/9, 4/7)
Mathtastic Level 1 - Numbers to 10
Add from largest, subtract by counting back
tracyashbridge.com

Jane had some T-shirts.
She got 2 more for her birthday.
Now she has 6 T-shirts.
How many did she have before her birthday?

Join – start unknown (3/7, 8/1)
Mathtastic Level 1 – Numbers to 10
Add from largest, subtract by counting back
tracyashbridge.com

Sam had 7 marbles.
He gave 4 to Mark.
How many marbles does Sam have now?

Separate – result unknown (8/3, 10/6)
Mathtastic Level 1 – Numbers to 10
Add from largest, subtract by counting back
tracyashbridge.com

Dave had 8 crayons.
He gave some to Jack.
Now he has 6.
How many did he give to Jack?

Separate – change unknown (9/5, 7/5)
Mathtastic Level 1 – Numbers to 10
Add from largest, subtract by counting back
tracyashbridge.com

Tim had some marbles.
He gave 7 to Jack.
Now Tim has 3 marbles.
How many marbles did Jack have at the start?

Separate – start unknown (6/2, 9/5)
Mathtastic Level 1 – Numbers to 10
Add from largest, subtract by counting back
tracyashbridge.com

Kim has 5 marbles.
Jan has 3 marbles.
How many marbles altogether?

Part – Part – Whole – whole unknown (8/2, 5/4)
Mathtastic Level 1 – Numbers to 10
Add from largest, subtract by counting back
tracyashbridge.com

There were 10 balloons.
4 blew away.
How many balloons are left?

Part – Part – Whole – part unknown (9/3, 6/2)
Mathtastic Level 1 – Numbers to 10
Add from largest, subtract by counting back
tracyashbridge.com

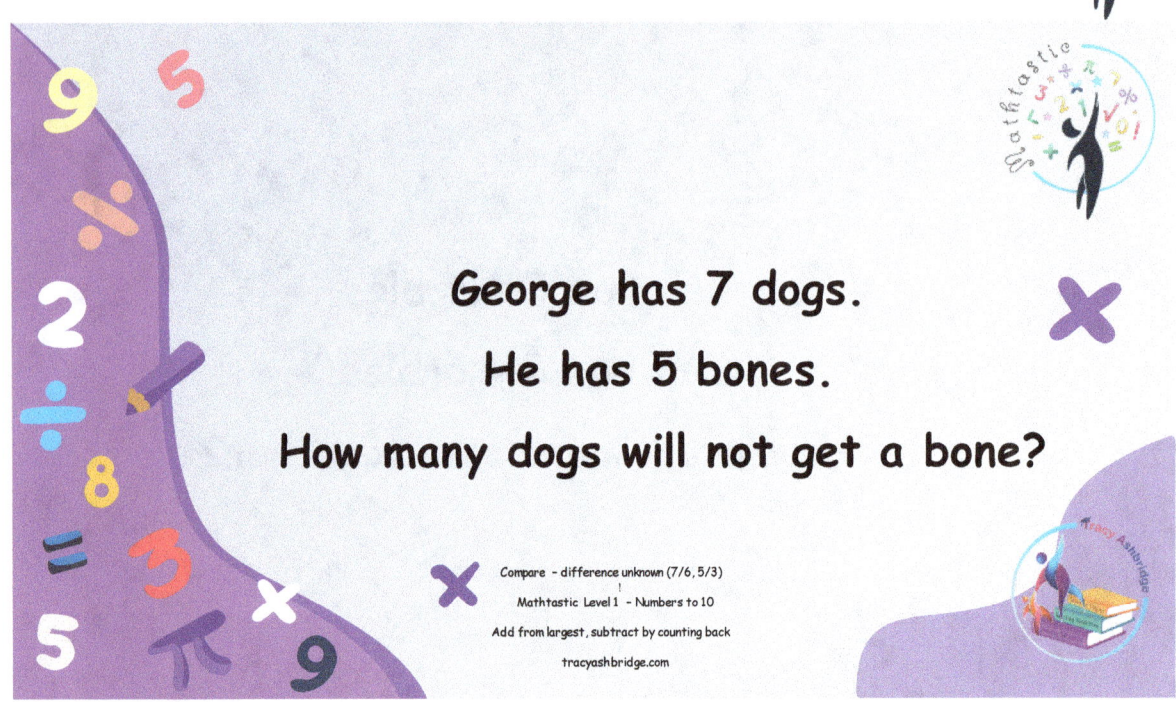

George has 7 dogs.

He has 5 bones.

How many dogs will not get a bone?

Compare – difference unknown (7/6, 5/3)
Mathtastic Level 1 – Numbers to 10
Add from largest, subtract by counting back
tracyashbridge.com

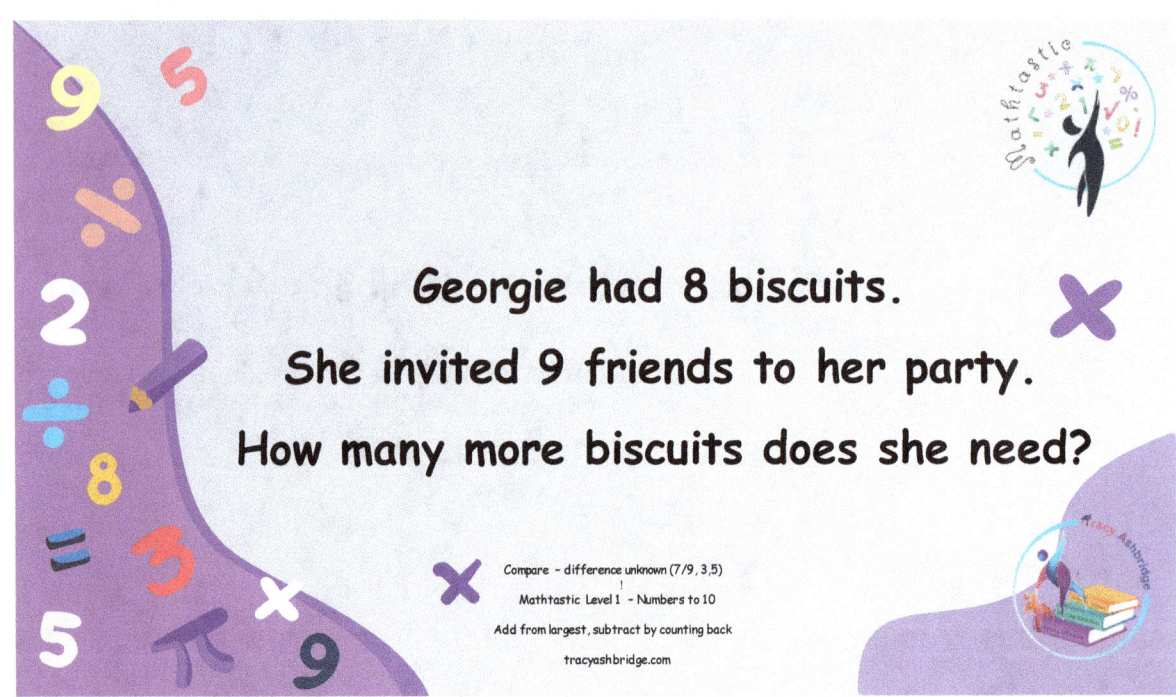

Georgie had 8 biscuits.

She invited 9 friends to her party.

How many more biscuits does she need?

Compare – difference unknown (7/9, 3,5)
Mathtastic Level 1 – Numbers to 10
Add from largest, subtract by counting back
tracyashbridge.com

Jenny has 4 guinea pigs.
Tracy has 3 more than Jenny.
How many guinea pigs does Tracy have?

Compare - referent unknown (7/2, 5/3)
Mathtastic Level 1 - Numbers to 10
Add from largest, subtract by counting back
tracyashbridge.com

Gemma had 3 biscuits.
Then she got 7 more.
How many biscuits does she have now?

Frank had 10 fish.
He gave 3 to his sister.
How many fish does Frank have now?

Jane had some fish.
She bought 5 more.
Now she has 10 fish.
How many did she buy?

Join – start unknown (4/10, 6/10)
Mathtastic Level 1 – Numbers to 10
Pairs to 10
tracyashbridge.com

Sam had 7 fish.
He sold 3.
How many does he have now?

Separate – result unknown (9/1, 8/2)
Mathtastic Level 1 – Numbers to 10
Pairs to 10
tracyashbridge.com

© Copyright 2022 Mathtastic: Tracy Ashbridge. All rights reserved

Dave had 10 mice.
Some escaped.
Now he has 6 mice.
How many escaped?

Tim had some biscuits.
His brother ate 3.
Now he only has 7.
How many did he have at the start?

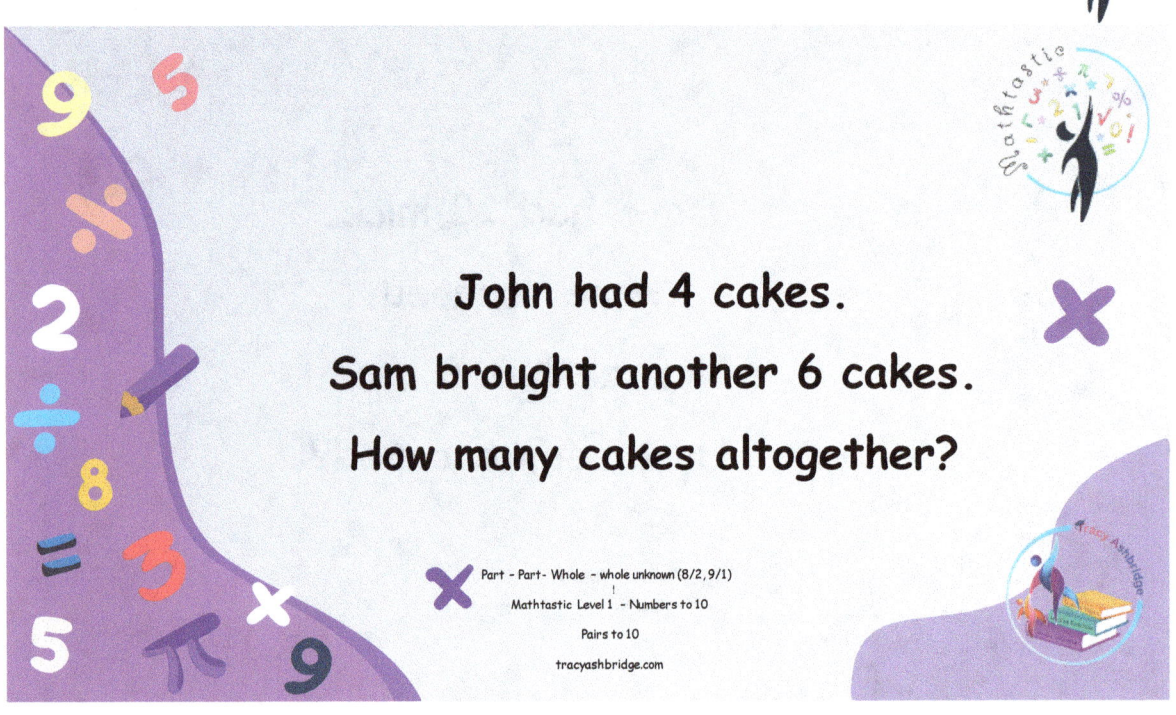

John had 4 cakes.

Sam brought another 6 cakes.

How many cakes altogether?

Part – Part- Whole – whole unknown (8/2, 9/1)
Mathtastic Level 1 – Numbers to 10
Pairs to 10
tracyashbridge.com

There were 10 cookies.
The mouse nibbled on 3.
How many were not nibbled?

Part – Part- Whole – part unknown (10/4, 10/9)
Mathtastic Level 1 – Numbers to 10
Pairs to 10
tracyashbridge.com

Georgie has 10 pens.
Some don't work.
6 do work.
How many don't work?

Compare - difference unknown (10/2, 10/7)
Mathtastic Level 1 - Numbers to 10
Pairs to 10
tracyashbridge.com

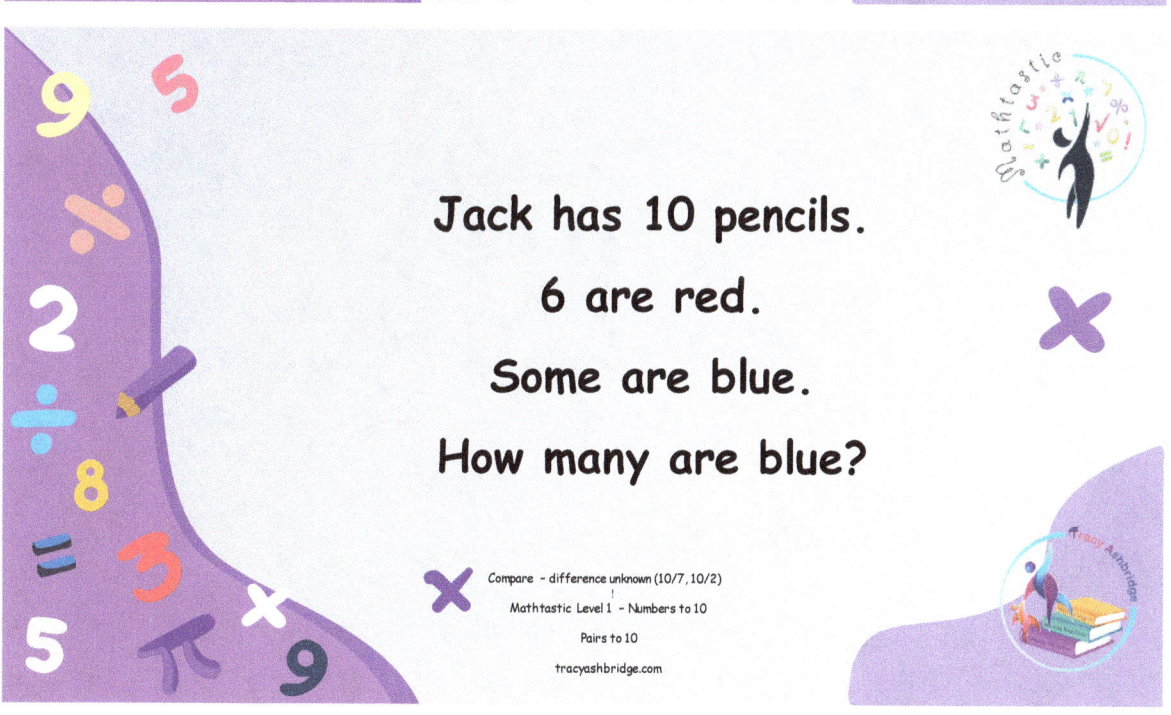

Jack has 10 pencils.
6 are red.
Some are blue.
How many are blue?

Compare - difference unknown (10/7, 10/2)
Mathtastic Level 1 - Numbers to 10
Pairs to 10
tracyashbridge.com

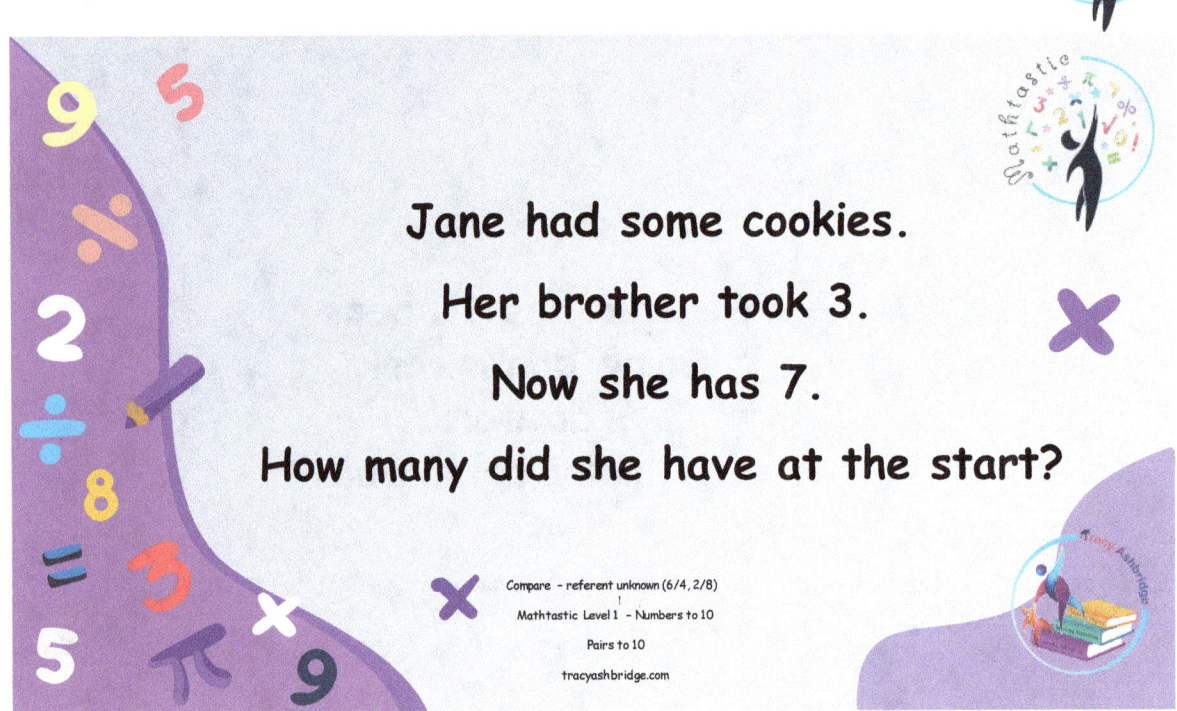

Jane had some cookies.
Her brother took 3.
Now she has 7.
How many did she have at the start?

Compare - referent unknown (6/4, 2/8)
Mathtastic Level 1 - Numbers to 10
Pairs to 10
tracyashbridge.com

Mathstastic

CGI Math Problems

Mathtastic Level 1 – Numbers to 10

add and subtract 5

tracyashbridge.com

Jane had some fish.
She bought 5 more.
Now she has 7 fish.
How many did she buy?

Join - start unknown (3/8, 2/7)
Mathtastic Level 1 - Numbers to 10
add and subtract 5
tracyashbridge.com

Sam had 5 fish.
He sold 2.
How many does he have now?

Separate - result unknown (5/3, 5/1)
Mathtastic Level 1 - Numbers to 10
add and subtract 5
tracyashbridge.com

Dave had 6 mice.
Some escaped.
Now he has 5 mice.
How many escaped?

Separate – change unknown (7/2, 8/5)
Mathtastic Level 1 – Numbers to 10
add and subtract 5
tracyashbridge.com

Tim had some biscuits.
His brother ate 2.
Now he only has 7.
How many did he have at the start?

Separate – start unknown (3/5, 5/2)
Mathtastic Level 1 – Numbers to 10
add and subtract 5
tracyashbridge.com

John had 5 cakes.
Sam brought another 3 cakes.
How many cakes altogether?

Part – Part– Whole – whole unknown (5/2, 1/5)
Mathtastic Level 1 – Numbers to 10
add and subtract 5
tracyashbridge.com

There were 8 cookies.
The mouse nibbled on 3.
How many were not nibbled?

Part – Part– Whole – part unknown (7/2, 5/3)
Mathtastic Level 1 – Numbers to 10
add and subtract 5
tracyashbridge.com

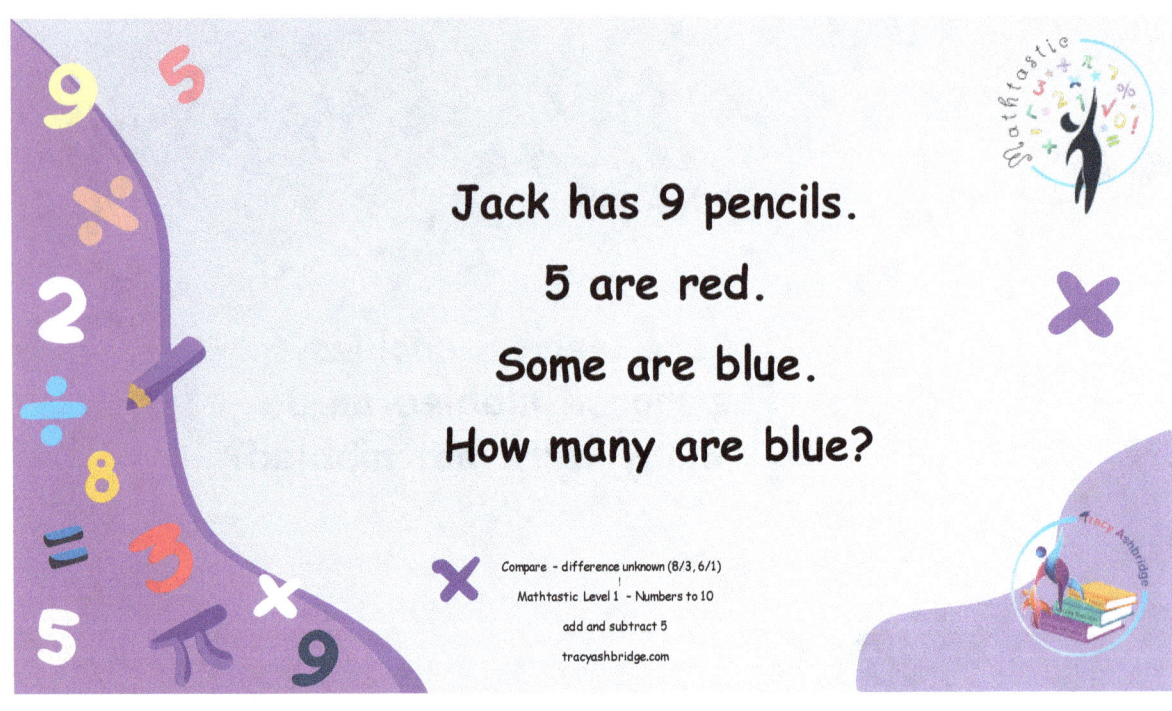

Jane had some cookies.
Her brother took 3.
Now she has 5.
How many did she have at the start?

Mathstastic
CGI Math Problems

Mathtastic Level 1 – Numbers to 10

Doubles to 10

tracyashbridge.com

Georgie had 3 crayons.

Sam gave her 3 more for her birthday.

How many crayons does she have now?

Join – result unknown (5/5, 4./4)
Mathtastic Level 1 – Numbers to 10
Doubles
tracyashbridge.com

Georgie had 4 crayons.

She got some more for her birthday.

Now she has 8 crayons.

How many crayons did she get for her birthday?

Join – change unknown (3/6, 2/4)
Mathtastic Level 1 – Numbers to 10
Doubles
tracyashbridge.com

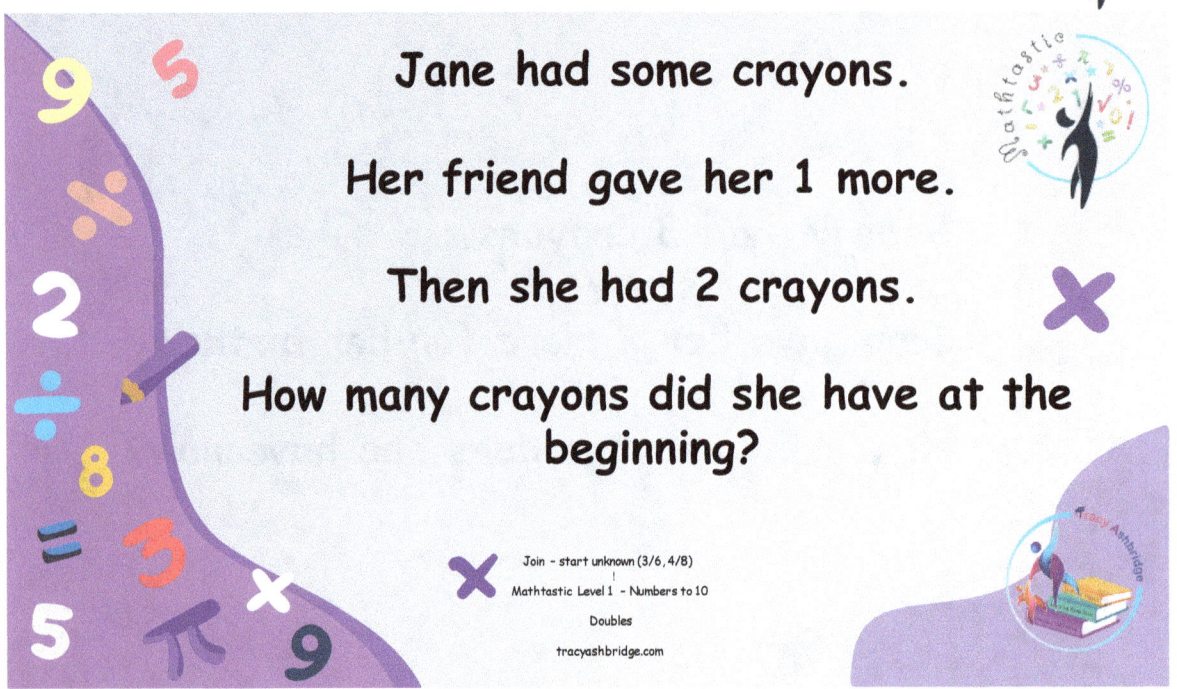

Jane had some crayons.

Her friend gave her 1 more.

Then she had 2 crayons.

How many crayons did she have at the beginning?

Join - start unknown (3/6, 4/8)
Mathtastic Level 1 - Numbers to 10
Doubles
tracyashbridge.com

Sam had 6 pencils.

He gave 3 to Sarah.

How many pencils does Sam have left?

Separate - result unknown (10/5, 4/2)
Mathtastic Level 1 - Numbers to 10
Doubles
tracyashbridge.com

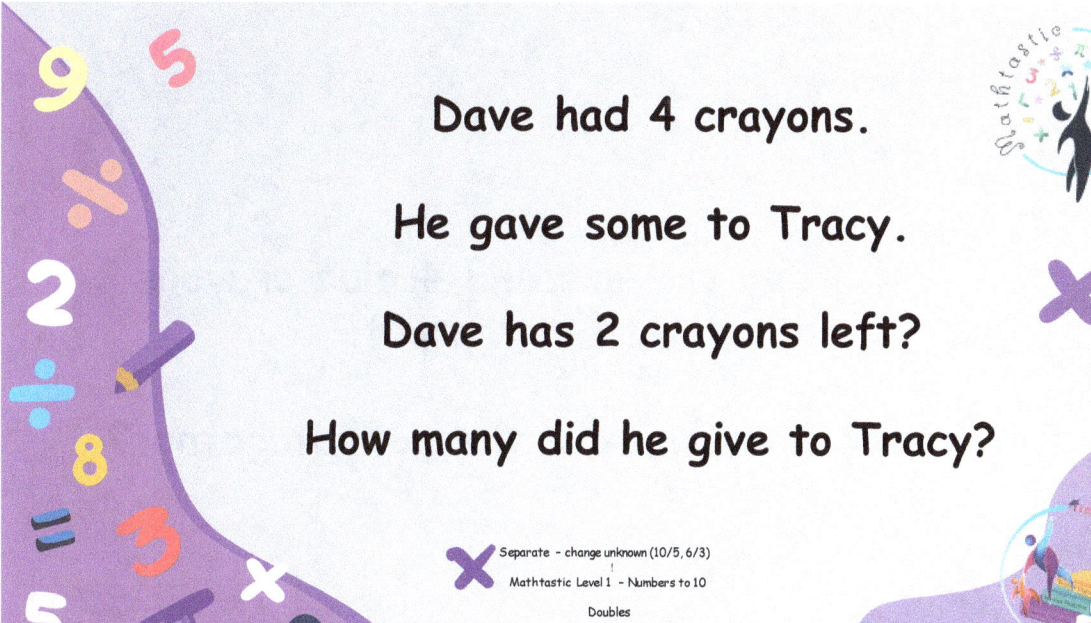

Dave had 4 crayons.

He gave some to Tracy.

Dave has 2 crayons left?

How many did he give to Tracy?

Separate – change unknown (10/5, 6/3)
Mathtastic Level 1 – Numbers to 10
Doubles
tracyashbridge.com

Tim had some crayons.

He gave 2 to Flynn.

Then he had 2 left.

How many crayons did Tim have to start with?

Separate – start unknown (1/1, 3/3)
Mathtastic Level 1 – Numbers to 10
Doubles
tracyashbridge.com

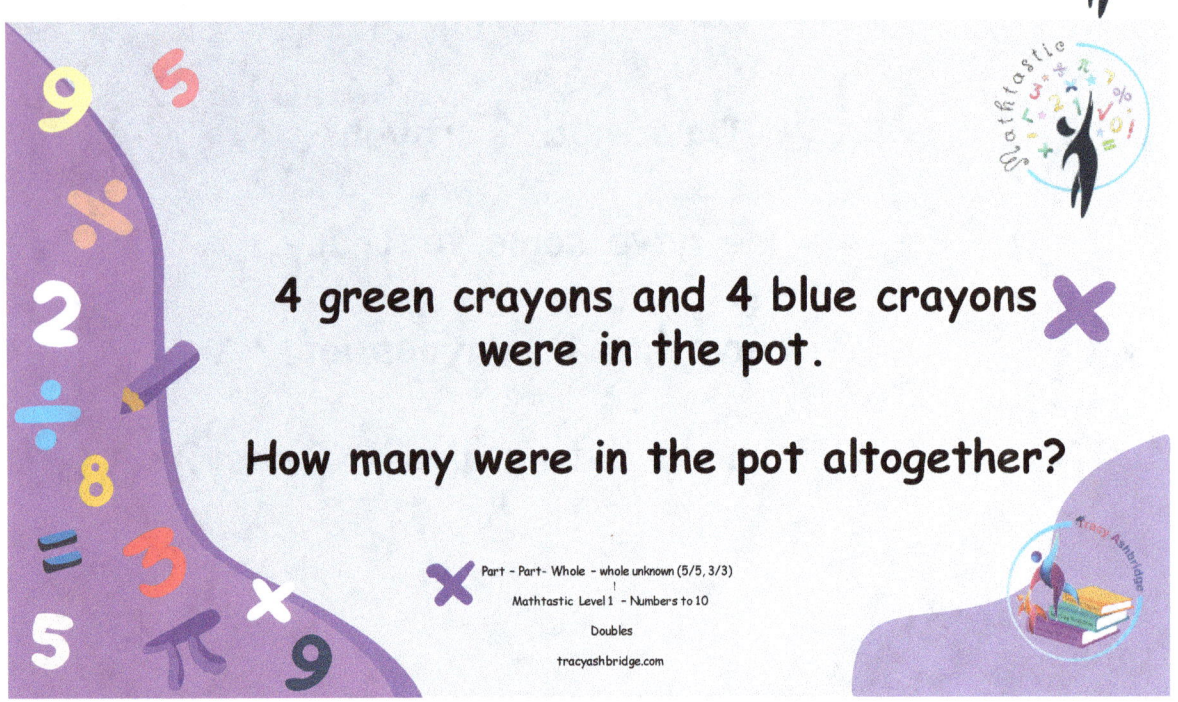

4 green crayons and 4 blue crayons were in the pot.

How many were in the pot altogether?

Part – Part– Whole – whole unknown (5/5, 3/3)
Mathtastic Level 1 – Numbers to 10
Doubles
tracyashbridge.com

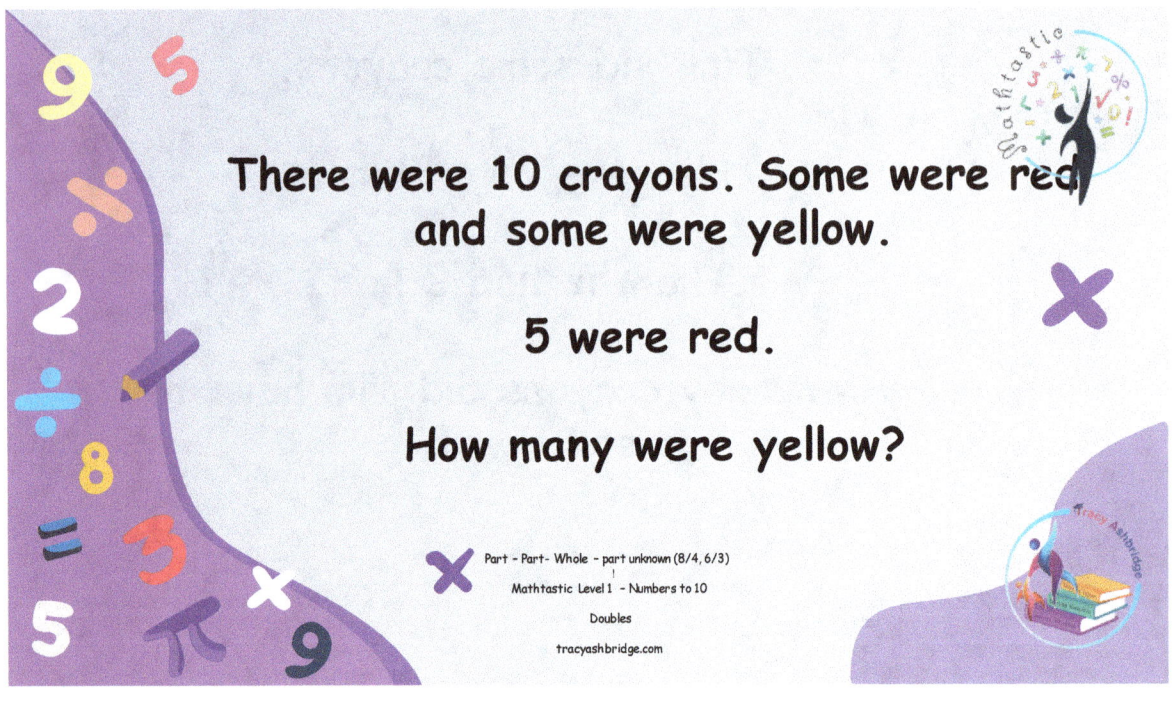

There were 10 crayons. Some were red and some were yellow.

5 were red.

How many were yellow?

Part – Part– Whole – part unknown (8/4, 6/3)
Mathtastic Level 1 – Numbers to 10
Doubles
tracyashbridge.com

© Copyright 2022 Mathtastic: Tracy Ashbridge. All rights reserved

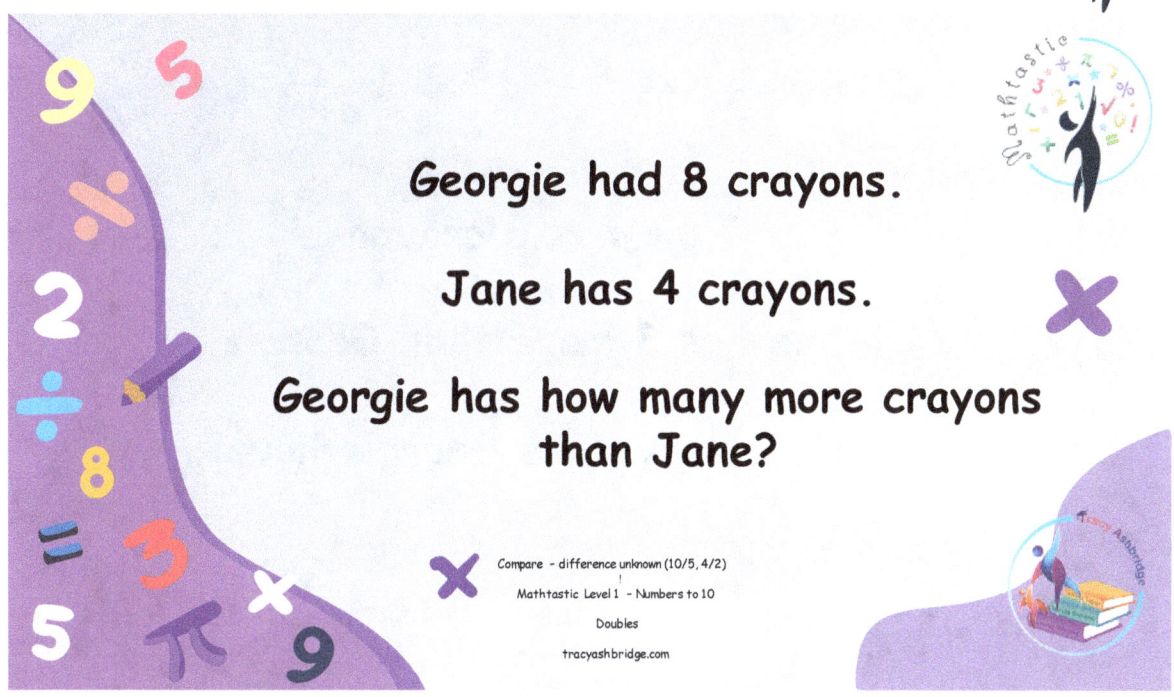

Georgie had 8 crayons.

Jane has 4 crayons.

Georgie has how many more crayons than Jane?

Compare - difference unknown (10/5, 4/2)
Mathtastic Level 1 - Numbers to 10
Doubles
tracyashbridge.com

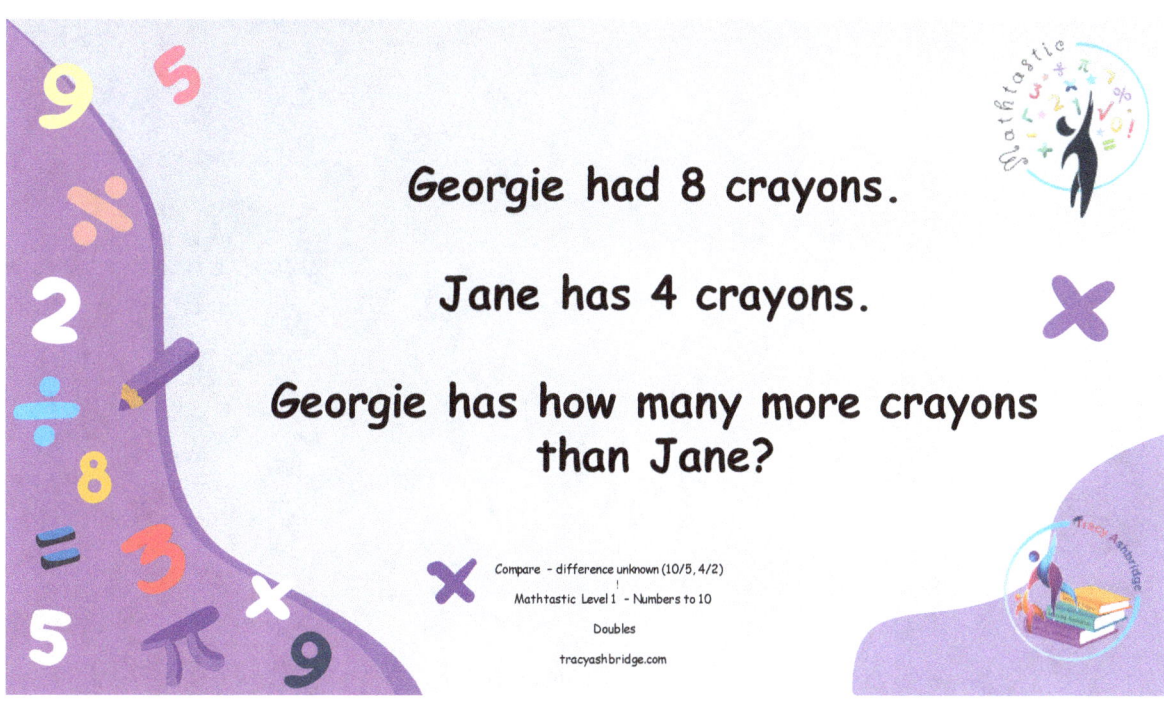

Georgie had 8 crayons.

Jane has 4 crayons.

Georgie has how many more crayons than Jane?

Compare - difference unknown (10/5, 4/2)
Mathtastic Level 1 - Numbers to 10
Doubles
tracyashbridge.com

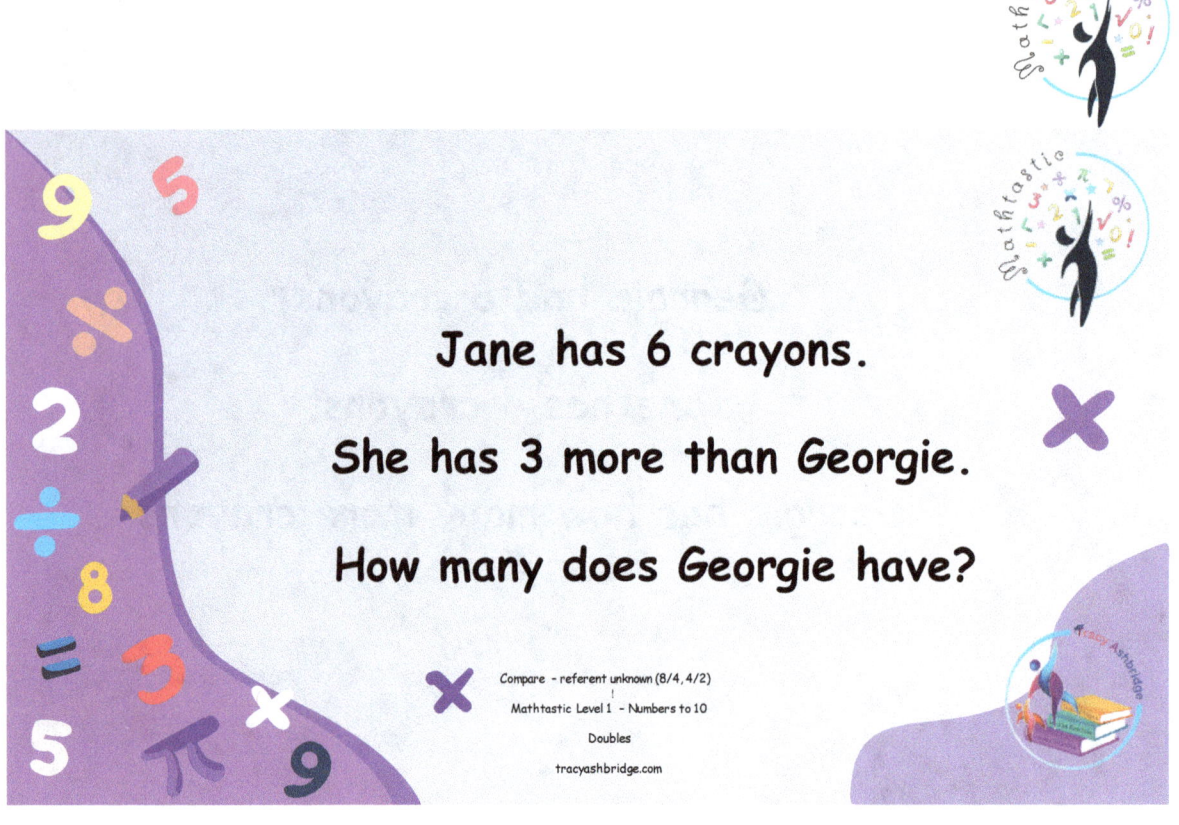

Mathstastic
CGI Math Problems

Mathtastic Level 1 - Numbers to 10

Near doubles

tracyashbridge.com

Emma had 3 biscuits.

Then she got 4 more.

How many biscuits does she have now?

Join – result unknown (4/5, 2/3)
Mathtastic Level 1 – Numbers to 10
Near doubles
tracyashbridge.com

Betty had 3 fish.

She gave 2 to his sister.

How many fish does Betty have now?

Join – change unknown (5/4, 4/3)
Mathtastic Level 1 – Numbers to 10
Near doubles
tracyashbridge.com

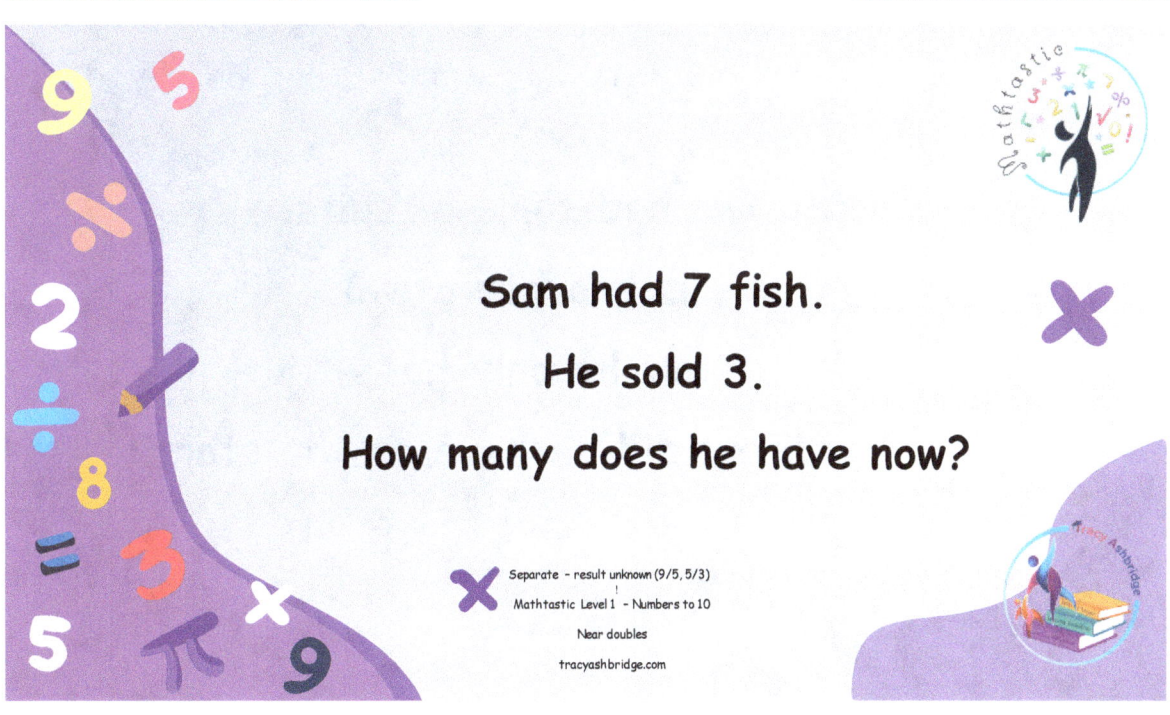

Dave had 9 mice.
Some escaped.
Now he has 5 mice.
How many escaped?

Separate – change unknown (7/4, 5/2)
Mathtastic Level 1 – Numbers to 10
Near doubles
tracyashbridge.com

Tim had some biscuits.
His brother ate 3.
Now he only has 4.
How many did he have at the start?

Separate – start unknown (5/4, 2/3)
Mathtastic Level 1 – Numbers to 10
Near doubles
tracyashbridge.com

John had 4 cakes.

Sam brought another 5 cakes.

How many cakes altogether?

Part – Part- Whole – whole unknown (2/3, 3/4)
Mathtastic Level 1 – Numbers to 10
Near doubles
tracyashbridge.com

There were 7 cookies.
The mouse nibbled on 3.
How many were not nibbled?

Part – Part- Whole – part unknown (9/4, 5/2)
Mathtastic Level 1 – Numbers to 10
Near doubles
tracyashbridge.com

© Copyright 2022 Mathtastic: Tracy Ashbridge. All rights reserved

Georgie has 9 pens.
Some don't work.
5 do work.
How many don't work?

Compare – difference unknown (7/3, 5/2)
Mathtastic Level 1 – Numbers to 10
Near doubles
tracyashbridge.com

Jack has 7 pencils.
4 are red.
Some are blue.
How many are blue?

Compare – difference unknown (9/4, 9/5)
Mathtastic Level 1 – Numbers to 10
Near doubles
tracyashbridge.com

Jane had some cookies.
Her brother took 3.
Now she has 4.
How many did she have at the start?

Compare – referent unknown (2/3, 4/5)
Mathtastic Level 1 – Numbers to 10
Near doubles
tracyashbridge.com

www.ingramcontent.com/pod-product-compliance
Lightning Source LLC
Chambersburg PA
CBHW080855010526
44107CB00057B/2588